North U.
Cruising &
Seamanship
Workbook

Bill Gladstone
John Rousmaniere

North U.
Cruising & Seamanship Workbook
Revised 3rd edition
 Bill Gladstone
 John Rousmaniere

Copyright © 2003 by Bill Gladstone

ISBN 0-9724361-7-0

Additional copies available through:
North U
29 High Field Lane
Madison CT 06443-2516

Printed in Hong Kong

The following illustrations appear with permission from:
Rousmaniere, John. *The Annapolis Book of Seamanship,* 3rd ed. (New York, NY: Simon and Schuster, 1999).

Page 3	Hove-To
Page 68	Deck and Interior Layout
Page 76	Quick Stop
Page 76	Reach-and-Reach
Page 83	Bowline
Page 84	Types of Rope
Page 84	Halyard Coil
Page 85	Buntline Hitch
Page 86	Fisherman's Bend
Page 87	Casting Off
Page 88	Line on a Winch
Page 88	Winch Overrides
Page 89	Anchor, Chain, Rode
Page 92	Anchor Scope
Page 95	Weighing Anchor
Page 95	Anchor Trip Line
Page 96	Dock Lines
Page 96	Docking
Page 97	Undocking, Stern Out
Page 97	Undocking, Bow Out
Page 98	Undocking from Slips
Page 98	Docking in Slips
Page 100	Deck and Interior Layout
Page 120	Winch Overrides

Gennaker and *Snuffer* are trademarks of North Sails Group, Inc.

Dacron is a trademark of DuPont, Inc.

Cruising and Seamanship Workbook
Contents

Chapter I – Sailing, Cruising and Seamanship 1
1. What is Cruising?   2. Seamanship
3. Cruising Boats   4. Experience Questionnaire

Chapter II – Upwind Sail Trim 7
1. Trim Theory in Brief   2. Speed, Pointing, Balance
3. Mainsail Trim and Controls   4. Genoa Trim and Controls
5. Helming, Tacking and Jibing   6. Moderate Air Trim   7. Light Air Trim
8. Heavy Air Trim   9. Performance Problems and Trim Solutions

Chapter III – Downwind Sail Trim 33
1. Reaching Trim   2. Running Trim   3. Gennakers and Spinnakers

Chapter IV – Heavy Weather Sailing 49
1. What is Heavy Weather?   2. Heavy Weather Forehandedness
3. Heavy Weather Sailing   4. Squall!
5. Storms   6. Alternate Storm Strategy
7. Misery and Danger

Chapter V– Safety 65
1. Risk   2. Formula for Disaster   3. Prepare the Boat
4. Prepare the Crew   5. Prevent Emergencies
6. Emergencies!   7. Equipment List

Chapter VI – Knots, Line and Gear 83
1. Rope and Line   2. Knots   3. Lines Under Load

Chapter VII – Anchoring and Docking 89
1. Anchoring   2. Docking

Chapter VIII – Sailing to a Destination 99
1. Welcome Aboard
2. Dead Reckoning, Coastal Piloting, Navigation Aids, Tides, Current, and Traffic
3. Sailing in Wind Shifts   4. Sailing at Night
5. Problems Along the Way   6. Conclusion

Chapter I – Sailing, Cruising and Seamanship

1. What is Cruising?
2. Seamanship
3. Cruising Boats
4. Experience Questionnaire

Sailing, Cruising and Seamanship

1. What is Cruising?

Putting to sea for fun – for other than commercial or military purpose – what an odd notion. Venturing into harm's way, inviting a confrontation with the elements, and to pay handsomely for the privilege, with no recompense save the pleasure of the endeavor itself – what pleasure is that?

That King Charles II of England, whose yacht is shown on the previous page, sailed for pleasure in the 1660s, might be taken to reflect more upon the burdens of his sovereign life ashore than on the attractions of the sea.

Yet here we are, those few of us who find some great reward in the challenges and beauty of cruising.

What is cruising? It's living in a boat while making your way to a destination. Racers sail *on* boats. Cruisers sail and live *in* boats. While power boaters are going somewhere, a cruising sailor on his boat is already there. You may go out for a weekend or a week or two and cruise along the coast, anchoring at night or tying up in marinas. Or you may make a long passage — say, to Bermuda or Hawaii. It's all cruising.

And why do we go? For the chance to commune with nature, for the challenge, for the camaraderie, for the chance to escape life ashore. Each of us, for our own reasons.

And what is required? Cruising draws on a broad mix of skills – sailing skills, for sure, but so much more. Navigation skills, mechanical, interpersonal and leadership skills. Part of the enduring appeal of cruising is the varied skills required to meet the new challenges each cruise presents.

2. Seamanship

The skills required for cruising fall under the title of Seamanship. Seamanship is not something that only expert cruising sailors do. All sailing involves seamanship. Seamanship means to sail safely and enjoyably in any kind of boat, weather, and situation. It also means anticipating problems — a state of mind called *forehandedness*. Yes, good seamanship includes such skills as how to tie knots, how to anchor, how to navigate without (and with) a GPS. It includes knowing the rules of the road, and having the skill to dock without causing major damage to the pier, your boat, *or your marriage*. We'll touch on all these skills.

One example of good seamanship is knowing how to stop the boat under sail — say, to make life more comfortable during a hard blow, or fix a piece of broken gear, or make a sandwich. That's the fine art of *Heaving-to*. Many sailors think that heav-

Hove-To, with jib backed, helm lashed down, and main trimmed hard. Not just a storm tactic, Heaving-to is a fine way to 'park' and take a break while under way.

ing-to is only for a big storm, but you can do it any time in most any boat. Here's how:

Trim the jib to windward, put the helm to leeward, and then adjust the main sheet until the boat sails herself very slowly on a close reach.

We'll cover *Heaving-to* in detail later...

Seamanship revolves around setting priorities. There are so many pieces of gear on a modern sailboat, so many gadgets to tweak, that many sailors are easily distracted from the most important priorities. What are those priorities? They can change from moment to moment.

Forehandedness

Looking ahead, knowing your options, and keeping them open, anticipating problems, balancing goals, risks, and abilities. All things which we will put under the broad heading of *Forehandedness*.

Sailing Priorities

Keep the boat balanced and on her lines. That means not having too much heel and weather helm (which makes the boat head up), and having no lee helm (which makes the boat head off). Extreme heel is slow and uncomfortable, and can be dangerous because you can lose control of the helm. If you have lots of weather helm then reduce power with techniques we'll discuss later.

Keep the boat moving ahead but in control. A boat is like a bicycle. If a bike isn't moving, it can't be turned. If a boat isn't moving, she can't be steered. You must have steerageway (enough speed to steer quickly).

Trim the sails correctly. The yarn or ribbon telltales on the jib luff and mainsail leech should stream aft most of the time. The sails' shape should be trimmed to match the wind and sea conditions. Proper sail trim can improve not just speed, but also a boat's motion in a seaway.

Know where the wind is from, and where it may shift. Wind indicators include the wind arrow at the masthead, telltales on the shrouds, wind on the water, flags, the wind on your face or hands. Where will it shift? Note the trends, know the forecast, and draw from your own experience.

Know your crew and your boat. Match your sailing ambitions to your crew, and your boat. Be careful not to push the limits of either.

A cruising boat with a full keel and attached rudder.

A more modern fin keel and spade rudder. We'll look more closely at keels in Chapter II, next.

3. Cruising Boats

What makes a boat a cruising boat? Beyond the hull, rig, and sails, most have engines. Most also have keels – at least monohulled cruising boats.

Cruising boats have other features.

Let's start with the hull and the interior. A good cruising boat has several comfortable bunks usable underway when heeled, as well as at anchor, plus a galley, a table for eating, a head (toilet and washroom), an engine, and stowage for clothing, food. And of course she has sails.

If she's heavy and slow, with a full keel, and her sails are relatively small, she is often called an "all-out cruiser" or simply "cruiser." As someone once said, "She'll go a looooong way, and take a looooong time getting there, too."

But if she has a turn of speed, with fin keel and spade rudder, and can do well in handicap racing, she's generally called a "cruiser-racer" or "racer-cruiser." That doesn't mean she's a bad cruiser – only a faster cruiser.

Much is made of the value of weight in cruising boats – some say heavier is better in storms – though overall design, not just weight, determines seakeeping ability in heavy weather. It can also be said that heavy boats get caught in more storms, as they are so slow in light to moderate winds.

Furthermore, weight makes everything harder. Sails have more load, and are harder to trim, anchors must be heavier, etc. Lighter is easier. But lighter boats also have a more active, jerkier motion in waves.

In truth, most coastal cruising is done in moderate winds, close enough to shore that we rarely sail in storms. But still, we must be prepared for rough weather.

There are trade-offs all around. Bigger means more room for gear (and people). Bigger is also faster - but usually harder and more expensive to sail.

A good cruising boat will have the ability to carry stores – food, and cruising gear, including a dinghy. Not just stowage, but a design which sails to her lines even when loaded.

Other common cruising features include some type of self steering, and a dodger.

Some people cruise in race boats. "That's not a cruising boat – it's too fast."

"But I'm cruising in it – doesn't that make it a cruising boat?"

In the end a cruising boat is any boat on which people cruise. For now, I'd suggest the boat you have...

Shown here are a range of cruising boats suitable for coastal sailing. Clockwise, from above: A moderate to heavy displacement sloop with a double-headsail rig (which makes shortening sail easy). A moderate displacement sloop under cruising spinnaker. A pocket cruiser which is easy to handle for a small crew, but also short onspace for crew and gear. Next is a cruising catamaran under spinnaker; catamarans offer plenty of room below. Finally, a light-displacement boat designed for racing that has taken her cruising family a long way.

4. Experience Questionnaire

Sailing Experience

A. How many years have you been cruising? _____ Done any racing? _____

B. What is the longest cruise you've taken? _____ days and _____ miles

C. What is the longest nonstop passage you've made under sail? _____ days and _____ miles

D. What are your cruising ambitions? Cruise for a week___ a month ___ a summer ___ Sail around the world____

E. Have you chartered? Where? _____

_____

Boat Type and Equipment

A. I own or have sailed a lot in: monohull catamaran trimaran

B. Boat's length over all _____ feet

C. Boat's rig: sloop cutter yawl ketch cat

D. Displacement: light moderate heavy

E. My sails include: _____

_____

Age of sails: _____

F. Anchor (s) carried on board:
____ Danforth type
____ Plow/ CQR
____ Bruce
____ Other _____

G. I own a:
__ safety harness
__ inflatable PFD,
__ combination harness/inflatable PFD

H. My boat has a:
____ VHF	____ Hot water
____ Cell phone	____ Dodger
____ GPS	____ Windlass
____ Radar	____ Refrigerator
____ Auto pilot	____ Shore Power
____ Knotmeter	____ Life raft
____ Wind instruments	
____ Wheel or ____ tiller	
____ Spinnaker - w. pole or no pole	
____ Dinghy - hard / soft/ engine	

Too small ... Too big.

Chapter II – Upwind Sail Trim

1. Upwind Trim Theory in Brief
2. Speed, Pointing, Balance
3. Mainsail Trim and Controls
4. Genoa Trim and Controls
5. Helming
6. Moderate Air Trim
7. Light Air Trim
8. Heavy Air Trim
9. Performance Problems and Trim Solutions

1. Theory of Lift

Sail trim is the effort to control and refine the flow of air around the sails. As we discuss sail trim, we'll describe flow from the *luff* to the *leech* in terms of the *depth*, or *draft*, in the sail. We'll be concerned with both the amount and the position of depth. Changes in trim will affect the lift and thus the power the sails generate.

While the existence of *lift* and related forces are generally recognized (planes fly and boats sail) the understanding of how lift is generated remains a point of contention. The old *slot affect* and *venturi* models have been debunked, and replaced with *Circulation Theory* and other models. Without getting into deep theory, let's take a look at what we (think we) know about sail shape, lift, and performance. We'll start with flow:

Flow

Air flows around a sail (or wing). The air flowing around the outside accelerates and travels *faster* than the air inside… Wait right there: Why must it flow faster around the outside?

Why Faster Around the Outside?

Air flowing against a concave shape (the inside) is slowed as it "piles up" while air flowing around the convex outside of a foil accelerates. This is an observed fact. Air flowing around the back is drawn into the space on the outside of the sail. Consider this: If the air flowing around the outside did not accelerate, a vacuum would form, which nature abhors. The air is drawn in to fill the vacuum – *it flows faster* to fill the vacuum. (And no, the air molecules that enter together do not have to leave together... a common misconception.) Telltales show the air flowing off the leech of the mainsail.

Stall

When the flow around the outside separates from the sail before it reaches the leech, the sail (or wing) is *stalled*. We see this on mainsails when the leech telltales disappear behind the leech.

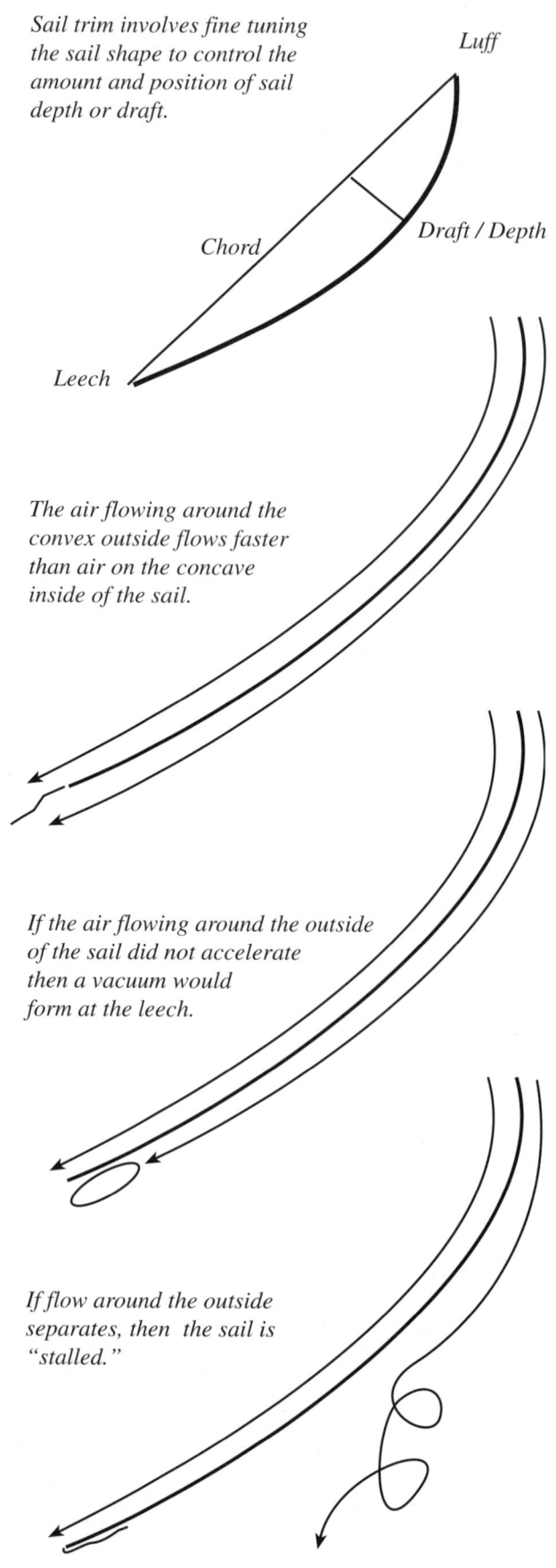

Sail trim involves fine tuning the sail shape to control the amount and position of sail depth or draft.

Luff

Chord

Draft / Depth

Leech

The air flowing around the convex outside flows faster than air on the concave inside of the sail.

If the air flowing around the outside of the sail did not accelerate then a vacuum would form at the leech.

If flow around the outside separates, then the sail is "stalled."

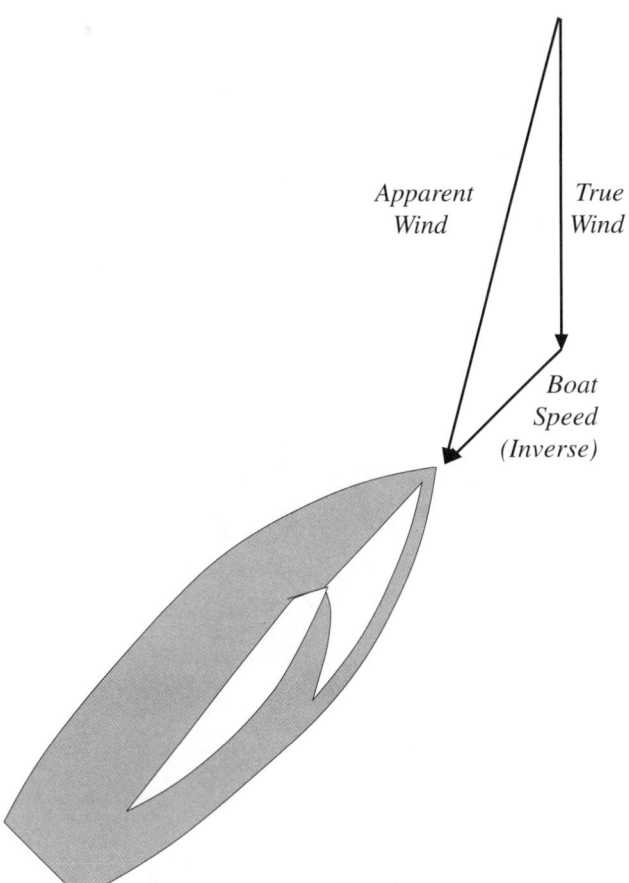

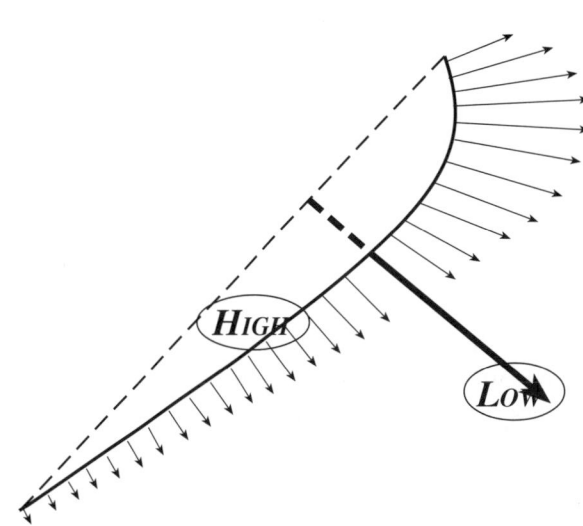

Slow flow on the inside creates high pressure, while the faster flow around the outside creates low pressure. This pressure difference is lift Advanced sail design encourages quick acceleration at the luff, for max lift forward. Speed differences are lower near the leech so less lift if generated there.

Apparent Wind

The boat's motion combines with the true wind to create our apparent wind. It is the apparent wind to which we trim, and which the sails use to create lift.

Forces of Lift

The faster moving air on the outside of the sail exerts less pressure on the sail than the slower moving air on the inside – that is *Bernoulli's Principle*. The relatively low pressure on the outside of the sail creates lift perpendicular to the chord of the sail.

When we put these sail lift forces on a boat we find a large, unwanted heeling force, and a relatively small forward force, as well as drag, or friction. One goal of trim is to improve the ratio of forward forces to heeling forces; another goal is to reduce drag.

Forces from the sails
1. Lifting force
2. Drag
3. Total force
4. Forward force
5. Heeling force

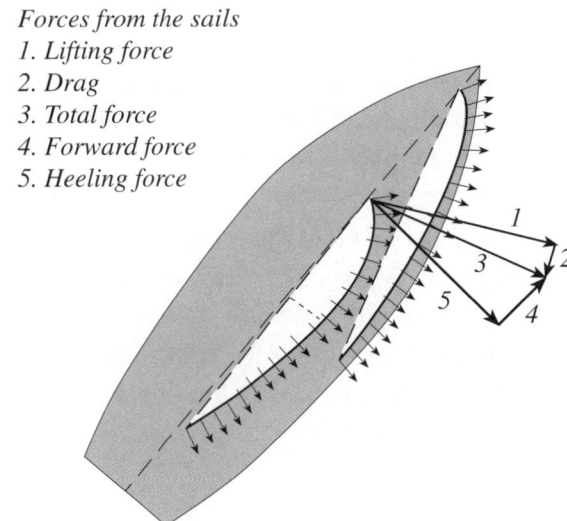

Main and Jib Working Together

The combined effect and interaction of the main and jib is a dangerous theoretical frontier. What is known is that the two sails work together to create a combined force greater than the sum of the forces the two sails could generate independently.

The flow of air is bent as it approaches the sail plan, putting the jib in a relative lift – which makes the jib more efficient – and the main in a relative header. We can see this in sail trim, as the jib is trimmed outboard, while the main is trimmed near the centerline. At the same time, the jib helps shape the flow of air around the main, which makes the main more efficient.

The two sails work together, with air flowing around the outside of the jib, and the inside of the main, as well as through the slot between the sails. We can see this flow on the telltales at the luff of the jib, and along the leech of the mainsail.

The Slot

Not all the air flows outside the jib or inside the main. Some flows through the slot, but not as much as you might imagine. The air which does flow through the slot flows onto the back of the main.

Add it Up

You can even view the main and jib as inside and outside sections of a single foil. No matter how you look at it, when you take the main and jib together we find a combined force which is mostly heeling force, with a small forward force.

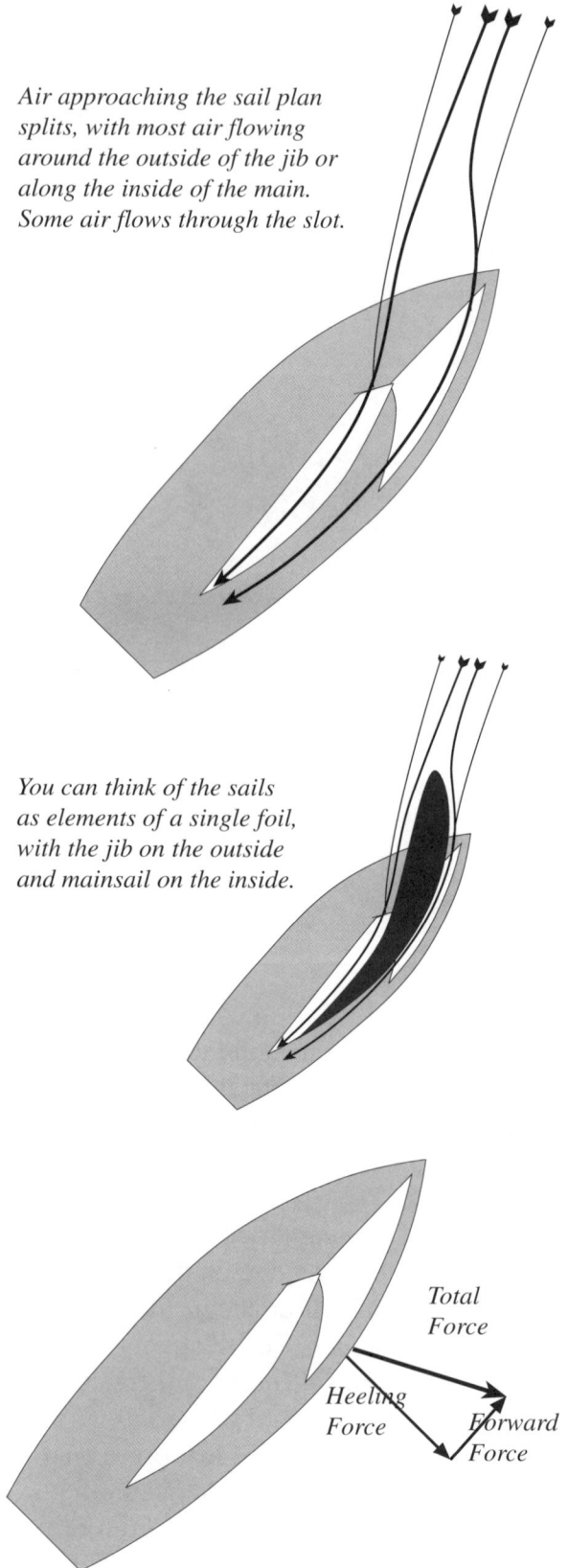

Air approaching the sail plan splits, with most air flowing around the outside of the jib or along the inside of the main. Some air flows through the slot.

You can think of the sails as elements of a single foil, with the jib on the outside and mainsail on the inside.

Total Force

Heeling Force

Forward Force

As we've seen, the total force from the sails can be broken down into a large heeling force and a small forward force.

Keel Lift

Were it not for the keel (or centerboard), the side force would be dominant; and we would not be able to sail upwind. Fortunately, the keel generates lift which nearly offsets the side force of the sails, so we can sail upwind (with only a few degrees of leeway).

"But wait a minute, how can the keel generate lift when it is symmetric?" I hear you ask.

"Ever see a plane fly upside-down?" I reply cleverly.

The issue here is *angle of attack*. While the keel is symmetric, the water does not hit it straight on; due to leeway the water hits the keel from a few degrees to leeward, and does not *see* a symmetric shape. It sees a foil with a long and a short side; and lift is generated perpendicular to the angle of attack.

Speed First

In order for the keel to generate lift it must first be moving through the water. You need speed first before you try to point. Look again at the forces on the boat: Only the keel takes you upwind. The sails push you downwind. The keel will take you upwind when you are moving fast. *Speed First*.

Keel Design

Given the critical importance of keel lift to upwind performance, we should take a brief look at keel design. While we won't pretend to be hydrodynamic experts, there are a few comments we feel qualified to make:

First, a fin keel provides better upwind performance than a full keel. Second, there is no substitute for depth. A deep fin keel will provide more lift than a shoal keel.

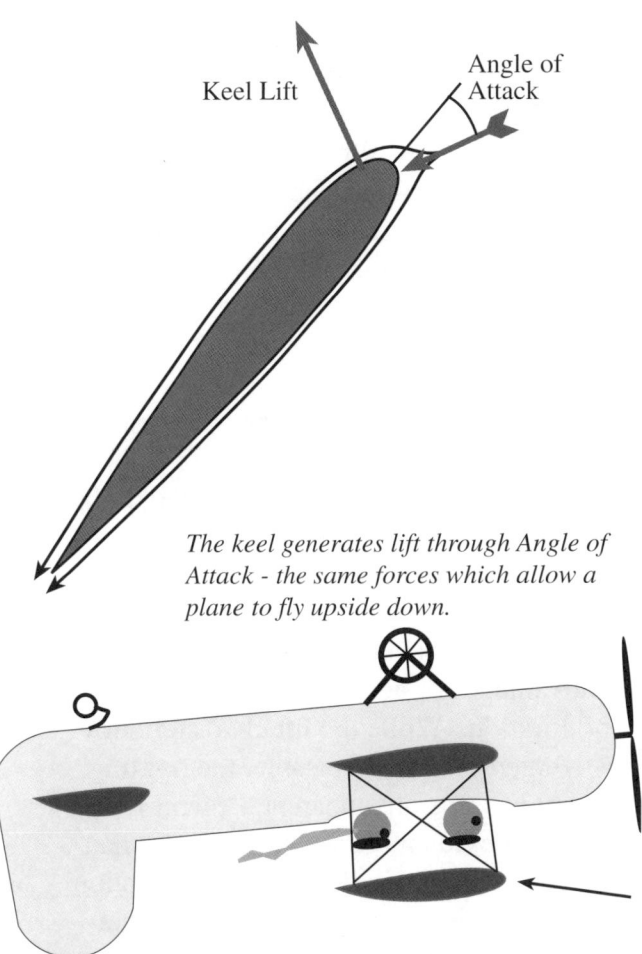

The keel generates lift through Angle of Attack - the same forces which allow a plane to fly upside down.

This fin keel is photographed at an angle of attack. A fin keel provides high lift and low drag for efficient upwind performance.

Third, a fair, smooth keel is more efficient than a rough or overgrown keel. Take care of your keel, and keep it clean.

Of course, keels do more than provide lift. They also provide stability. The more weight you carry low in the keel, the greater the righting force (righting moment) the keel will have. Increased stability increases sail carrying capability, and thus speed, at least upwind and reaching.

Keels also create drag. The more surface area the keel has, the more drag it creates. The drag of a full keel hinders light air performance. A fouled keel (and bottom) robs speed. Keep your keel and bottom clean.

Winged keels purport to provide some of the advantages of depth in a shallower configuration. While the lift characteristics of a winged keel are debatable, the righting moment benefits are apparent. A winged keel will provide more weight at depth than a fin keel of the same depth. Another option for designers is to place a bulb, rather than wings, at the base of a fin keel.

While there may be no substitute for depth in upwind performance, when sailing in six feet of water the advantages of a five foot draft over a seven foot draft are dramatically apparent.

The Combined Force of Keel and Sails

The combined forces of the keel and sails drive us forward. Note that only a very small fraction of the forces generated are actually translated into forward force. Most of our trimming and fine tuning effort is directed at improving this mix of useful (forward) and useless (heeling) forces. Even a slight improvement in the mix can make a big difference in performance. We'll look at details of upwind trim next.

This large cruising yacht sports a shoal draft keel with a centerboard. The centerboard is lowered to provide lift when sailing upwind. Though the center board does not improve righting moment, the ability to lift the board has clear advantages in shoal water.

A winged keel concentrates weight low for increased stability. One hazard of wing keeled boats is that depth increases as they heel, so when they run aground heeling will not lift them off.

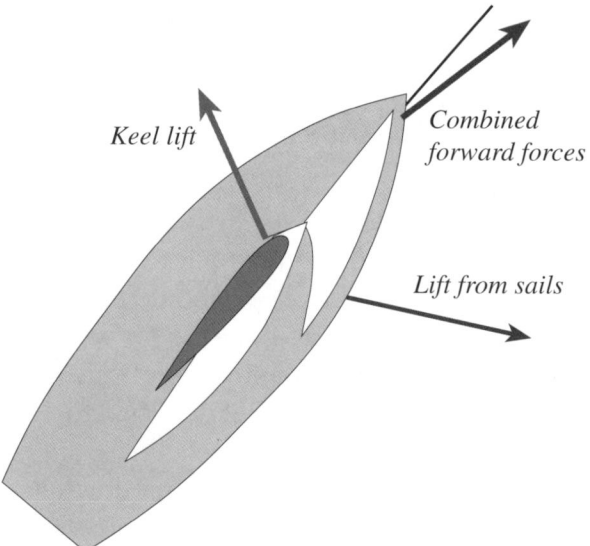

Keel lift

Combined forward forces

Lift from sails

The combined lifting forces of the sails and keel allow us to sail upwind with only a few degrees of leeway.

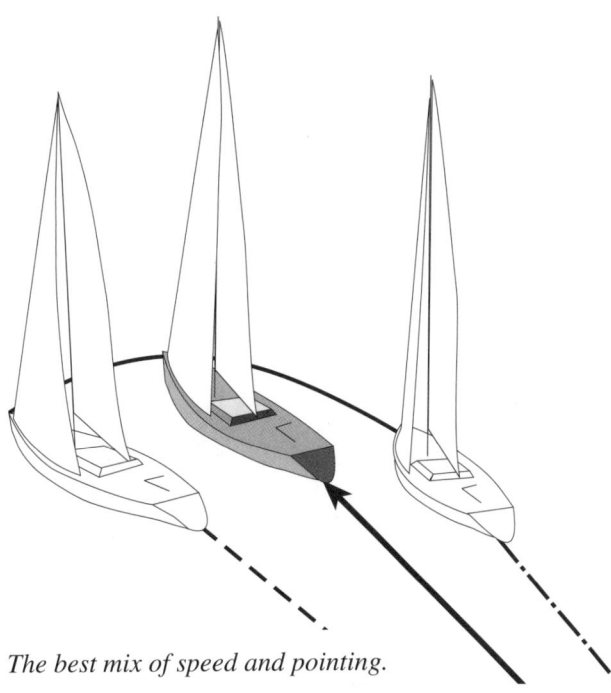

The best mix of speed and pointing.

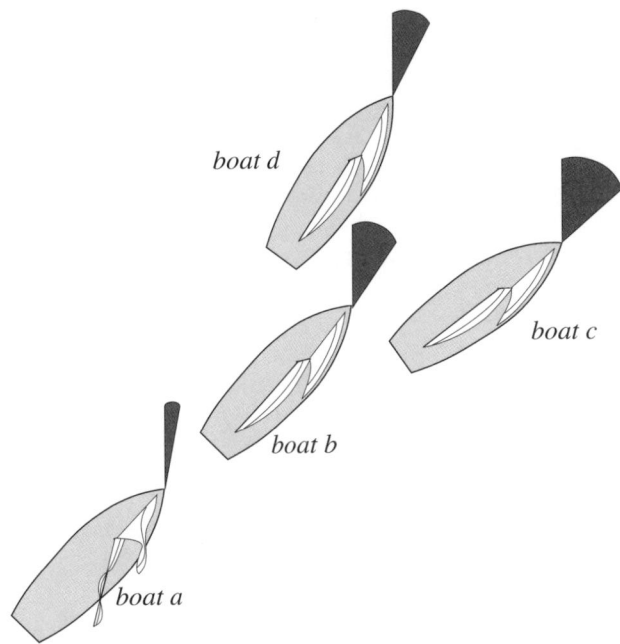

Angle of attack *is the first source of sail power. With sails luffing, angle of attack is zero, and there is no power (boat a). Trimming in widens the angle of attack, and increases power (boat b). Bearing off also widens angle of attack, increasing power (boat c), while heading up reduces both angle of attack and power (boat d).*

2. Speed, Pointing, Balance

Our primary goal when sailing upwind is to find the mix of speed and pointing which takes us upwind most quickly. To achieve the best performance we need to optimize the trim of each sail, and also the trim of the two sails together.

In this chapter we will look *first* at optimum upwind sail shape for mainsails and jibs. As we discuss each sail, we will look at the three sources of power in each sail: angle of attack, depth, and twist.

Only *after* determining optimum shape will we cover the sail controls, including our primary controls – the mainsheet and jib sheet, and secondary controls – such as halyards, cunningham, outhaul, backstay, traveler, boom vang, and jib leads, used to achieve that shape.

From there, we will look at helming, including how best to set the boat up for self steering. Next will come trim in various sailing conditions, starting with moderate air trim as a baseline, and including a look at light air and heavy air techniques.

Finally, we will look at a series of performance problems, and trim solutions.

3. Mainsail Trim

The mainsail, like any sail, has three sources of power. They are:
> Angle of Attack
> Depth
> Twist

Definitions:

Angle of attack is the angle at which the wind hits the sail. A wider angle brings more power. At zero angle of attack sails are luffing. As we trim in sails, or bear away from the wind, angle of attack and power increase.

Depth is the amount of curve in the sail. We measure depth from luff to leech. Sail depth is described as a proportion of the distance from luff to leech. A *deep* sail is sometimes referred to as a *full* sail, whereas a *flat* sail is, well, a flat sail. In addition to controlling the *amount* of depth in the sail we can also control the *position* of that

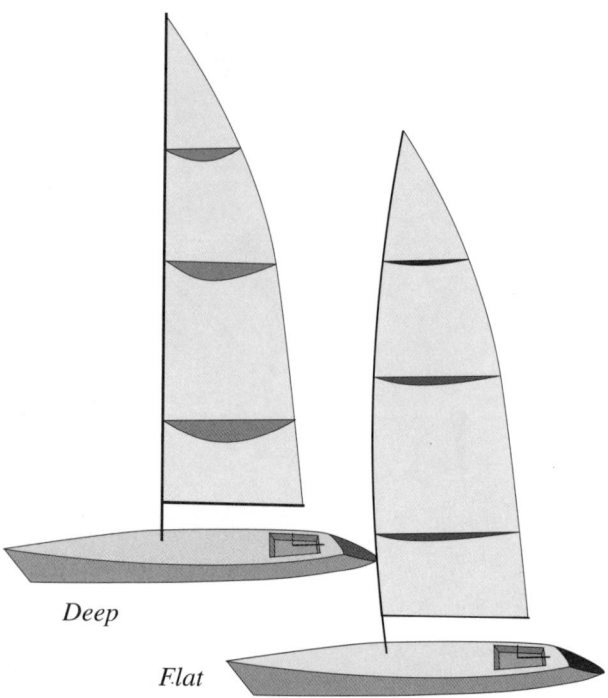

Deep

Flat

Sail Depth is the second source of sail power. Deeper sails are more powerful.

Lots of twist,
little power

Little twist,
lots of power

Twist is the third source of sail power. Adding twist reduces power by spilling wind from the upper portion of the sail. Changes in twist have a big impact on heeling forces and pointing ability.

deepest part. We'll describe how when we review the sail controls, later.

The third component of power is *Twist*. Twist describes the change in angle of attack from the bottom to the top of the sail. A sail with lots of twist is open at the upper leech, spilling power aloft. A sail with little twist has a closed upper leech – the upper leech is nearly parallel to the lower leech, and the sail has power aloft.

Sail power is the sum of power derived from each source. Proper trim means getting not just the correct *total* power, but also the correct *mix* of power.

Optimum Shape and Power

Angle of Attack is the first component of mainsail power. In most conditions, we trim the sail to put the boom on the centerline. When we are *overpowered* – indicated by too much heel and weather helm – we can reduce power by easing the boom down, below the center line, or by heading up slightly.

Mainsail Depth is set to match the conditions. Assuming a well designed sail - *in good condition* - we'll start out trimming the sail to its designed depth. From there, add depth if the boat needs more punch in light air or chop, and reduce depth to reduce power when necessary.

Twist is the third component of power. Talking about twist can be a little confusing, as *reducing* twist *adds* power and pointing ability. We can judge twist in most conditions through the telltales along the leech of the main. For starters, reduce twist to the point of *nearly* stalling the upper leech telltales. This will be near optimum pointing and power. From there, experiment: If the boat is overpowered, or heeling too much, then add twist, and spill some power. If you'd like to point slightly higher, reduce twist further; overtrim to the point of partially stalling the telltales. Ease out again if you feel you are losing speed.

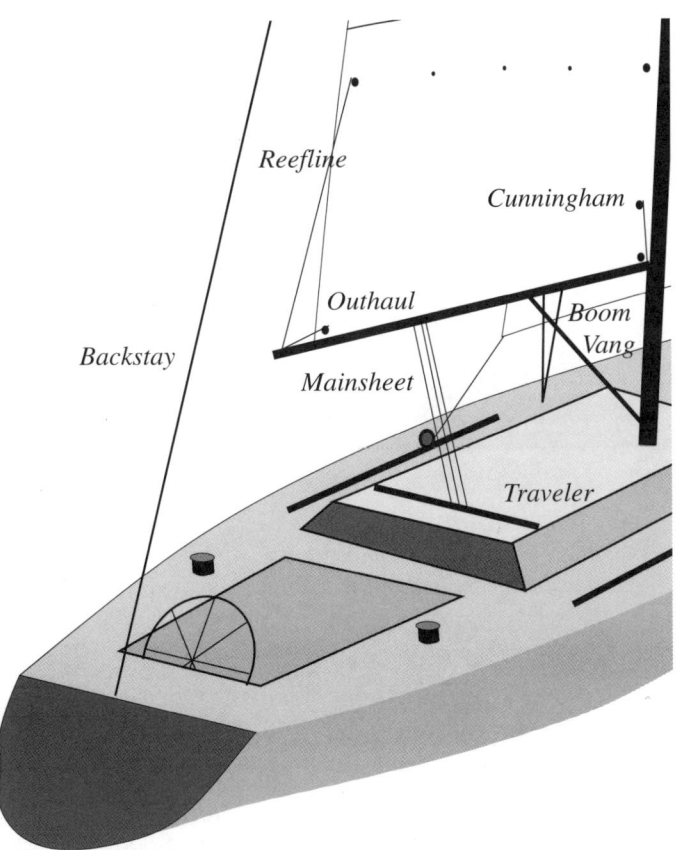

We adjust for mainsail trim and shape with a number of controls. The primary control is the mainsheet.

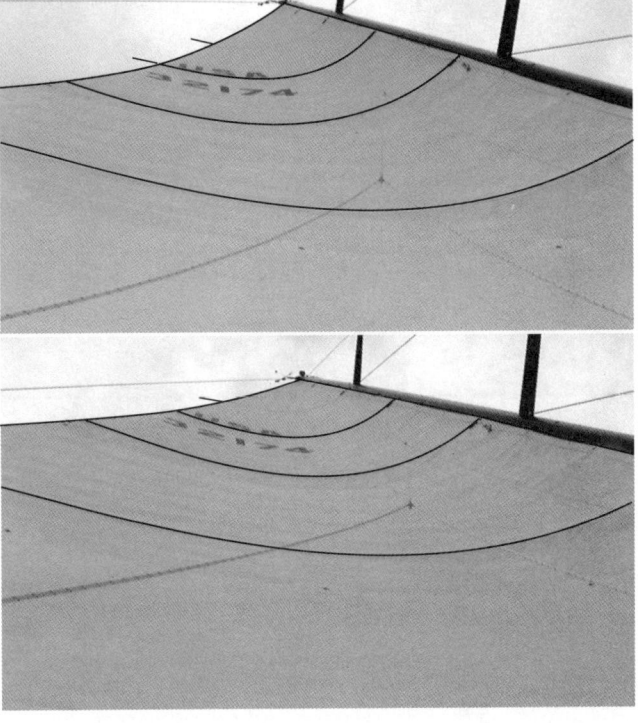

The mainsheet fine tunes mainsail twist. A more twisted, or open shape, as shown at top, provides less power. Trimming harder reduces twist and closes the leech, adding power, as in the lower photo. For best upwind performance trim so the upper mainsail leech parallels the boom while the upper leech telltales flow with an occasional stall.

Mainsail Controls

The mainsail has a wide variety of controls. The most important is (surprise!) the mainsheet. Other controls include the boom vang, mast bend, outhaul, traveler, and halyard / cunningham. If your boat is missing some of these secondary controls, then you simply use the controls you do have to do what you can.

Mainsheet

The mainsheet impacts every component of mainsail power. Initially, as we trim the mainsheet from a reach to close hauled, the primary impact is on angle of attack. As the boom nears the centerline, the primary impact of sheet trim is on twist – twist is reduced as the sheet is trimmed and the leech is tensioned. The mainsheet also impacts sail depth, but to a lesser degree.

For best upwind performance trim the mainsheet so the upper leech or upper batten of the main is parallel to the boom, and so the leech telltales are flowing, with only an occasional stall.

From there, experiment – try extra trim for extra pointing, with the telltales stalled more than half the time, or ease a little, and see if you don't add a boost of speed without any loss of pointing.

Changing twist with the mainsheet will impact pointing ability and also helm balance. Sheet harder to point higher. Ease, and add twist, if the boat is overpowered, slow, or difficult to steer.

As we move on to secondary controls, remember to always recheck mainsheet trim after any secondary adjustment.

This mainsail leech shows good closehauled trim. If the sheet were trimmed harder, the upper leech would close and flow would stall.

The traveler car is pulled above centerline in order to position the boom on the centerline for optimum moderate air upwind performance.

Traveler

The traveler positions the boom, changing angle of attack. Except when overpowered, adjust the traveler to keep the boom nearly centered. In overpowering conditions, ease the traveler down to reduce power and relieve weather helm.

Playing the traveler is particularly effective in puffy conditions. Once the overall sail shape is set for the prevailing conditions, play the traveler to quickly dump power in gusts.

On a boat that lacks a traveler or has one that's hard to adjust, ease the mainsheet to quickly reduce power.

Helm

Steering is another control of mainsail power. Heading up reduces power, while falling off widens the angle of attack, and adds power.

The helm also tells us if we are out of trim. While a slight weather helm is desirable, too much weather helm tells us we ought to reduce power.

Boom Vang

The boom vang is primarily an offwind control. Upwind, take the slack out to help control twist. In light air be careful to leave the vang eased.

Mast Bend

After the mainsheet, control over mast bend is the second most powerful controller of mainsail shape.

Mast bend flattens the mainsail by increasing the distance from luff to leech. The biggest impact is in the middle and upper portions of the sail. Adding mast bend reduces mainsail power.

As a secondary affect, mast bend also adds twist, by shortening the distance from head to clew.

Mast bend flattens the mainsail, reducing power, and also reducing drag. When the boat is fully powered, adding mast bend can increase speed by reducing drag – *but only when fully powered.* When overpowered, adding mast bend increases speed by reducing heel and weather helm, and the drag associated with them.

Outhaul

The outhaul controls depth in the lower portion of the mainsail. Pull the outhaul tighter as the wind builds, and ease it off for extra power in light air or chop. Don't get carried away - ease just enough to round the foot of the sail.

Halyard and Cunningham

The halyard and cunningham control luff tension, and through it, draft position - that

A cascading mainsheet system provides extra power for fine tuning main sail trim. The fine tune control is attached to what was the dead end of the mainsheet. At left, the fine tune is eased, at right, it is trimmed hard.

This sequence of photos shows the impact of mainsheet trim. As the sheet is trimmed twist is reduced, and power is added.. Trim harder to add power and point higher. Ease, and add twist for less power and a wider steering groove.
The top photo shows lots of twist - good trim for lighter air or choppy conditions. This shape would work well to spill power in heavy air as well.
The second photo shows little twist. This would be good trim for smooth water, moderate air sailing.
The main is overtrimmed when the leech closes and telltales stall, as in the bottom photo.

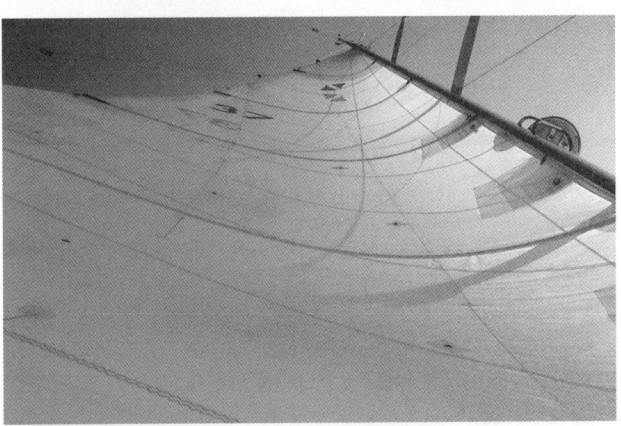

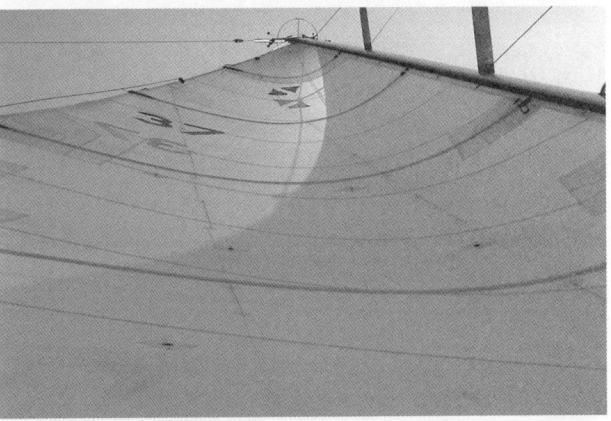

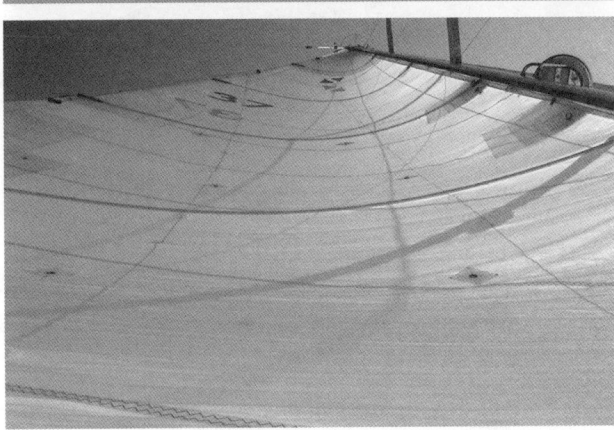

is, the position of the deepest part of the sail. We want to hold the draft in its designed position, just forward of the middle of the sail. We do this by tightening the halyard or cunningham as the wind builds. Use the halyard first, and when at full hoist, use the cunningham.

When you bend the mast the draft will move aft, so add luff tension as you add mast bend; and ease luff tension as the mast is straightened.

A Question

The question arises: In what order should we reduce power? As the breeze builds, should we reduce power through angle of attack, depth, or twist?

An excellent question, and one we will address in some detail in the heavy air trim section coming up later in this chapter....

But first, on to jibs and genoas.

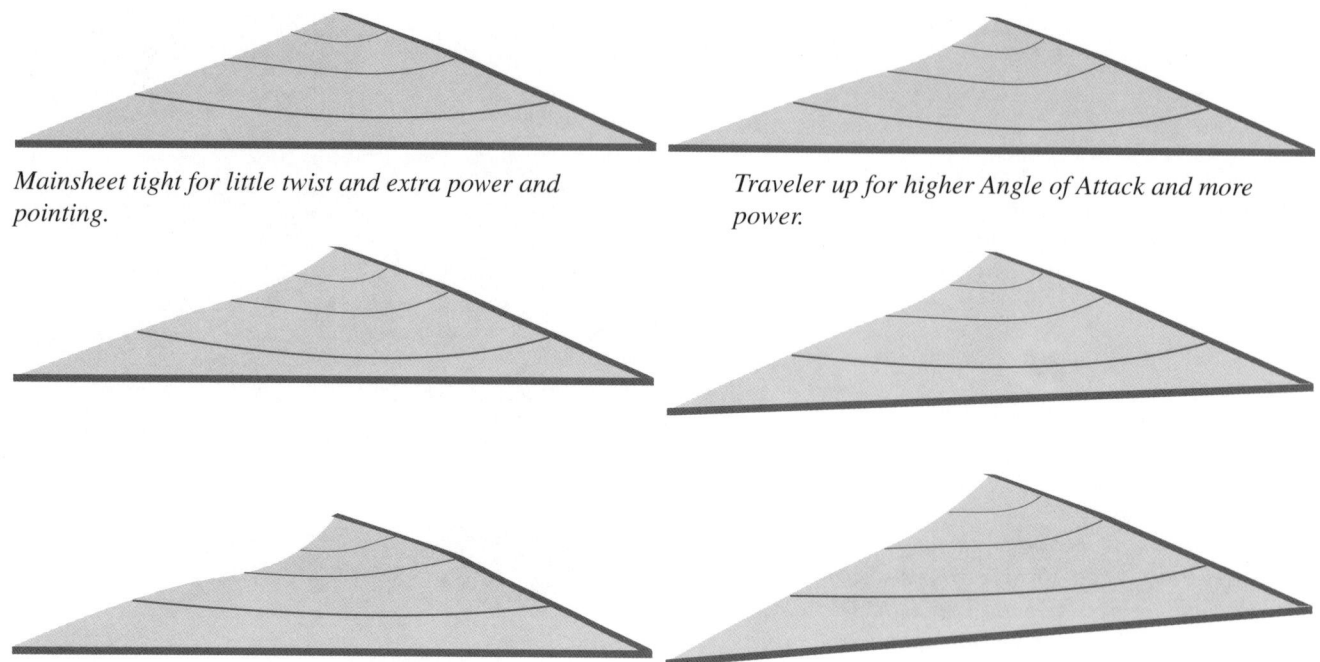

Mainsheet tight for little twist and extra power and pointing.

Traveler up for higher Angle of Attack and more power.

Mainsheet eased to add twist and spill power.

Traveler down reduces angle of attack and power

Mainsail trim and controls: Mainsail shapes can be fine tuned to suit the sailing conditions. These four sets of illustrations show changes in shape with each control. For optimum performance you can adjust trim constantly, or after the next snack or between chapters in your book or after a short nap or...

Mast straight leaves the sail deep and powerful.

Little luff tension allows the draft to move aft.

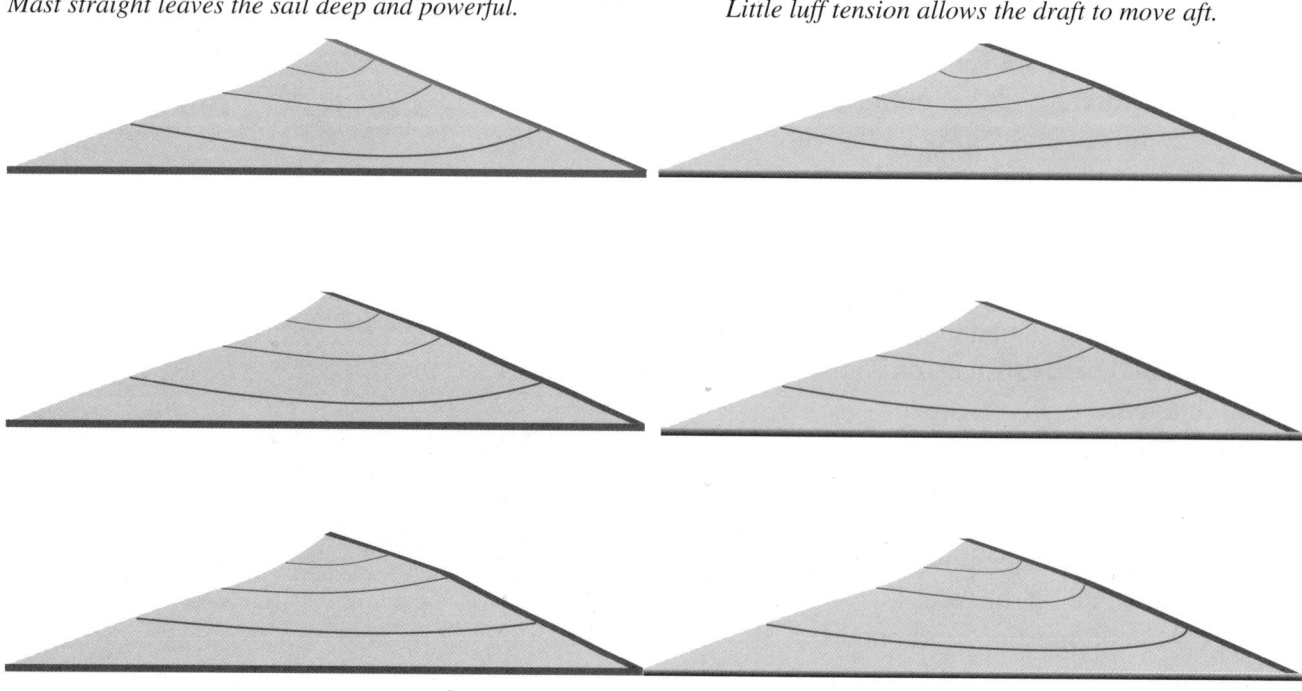

Mast bend makes the sail flat and depowered.

Luff tension added with halyard or cunningham pulls the draft forward.

North U. Cruising Workbook

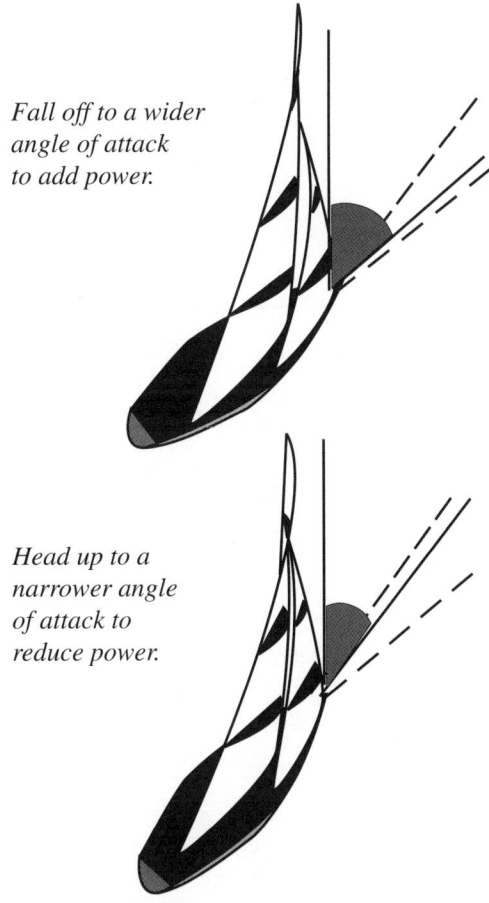

Fall off to a wider angle of attack to add power.

Head up to a narrower angle of attack to reduce power.

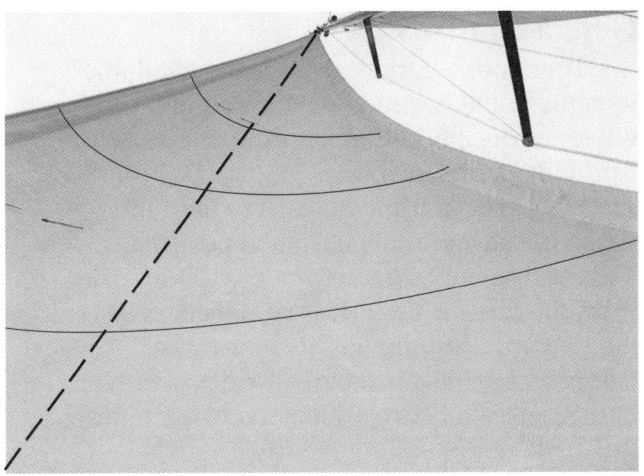

Initial genoa trim will create a sail shape like this, with moderate depth, and twist to match the shape of the main. The draft is positioned forward and there is sufficient entry shape to create a forgiving steering groove. The hook in the UV cover along the leech is of no consequence.

4. Genoa Trim

As with the main, there are three sources of genoa power: Angle of attack, sail depth, and twist. Our goal is first to get the correct total power, and second, to get the correct mix of power to suit conditions. The three sources of power are:

Angle of Attack

The genoa derives power first through angle of attack. Trim the sail in, and you add power. Ease the sail out and you reduce power. Once the sail is sheeted in, then the primary control of angle of attack is the helm. Fall off to add power, and head up to reduce power.

Angle of attack is increased by trimming the sheet or by falling off.

Depth

Deeper sails generate more power. Flat sail shapes generate less power (and less drag). Genoa depth is adjusted through a variety of controls, including headstay sag, lead position, and sheet trim.

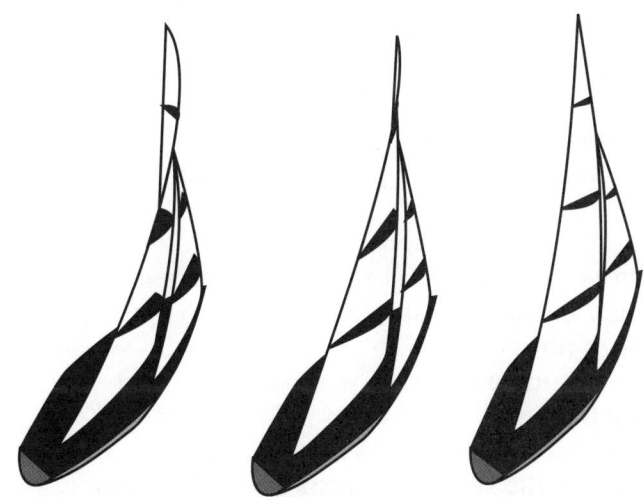

As sheets are trimmed twist is reduced, and power is increased.

Twist

A closed leech generates more power. A twisted, or open leech, spills power. Genoa twist is controlled through lead position and sheet trim.

Initially, the sheet's primary impact is on angle of attack, pulling the sail in. As the sail nears full trim, the sheet pulls the clew down (more than in), and the primary impact of trim is to change twist.

Total Power, Mix of Power

The objective is to achieve the correct total power in the sail, and also the correct mix of power from each source.

Genoa controls

There are a variety of controls available to achieve the desired amount and mix of power. Your jib will have some or all or the following sail controls:

Halyard: Set the halyard to hold the shape of the sail in its designed position. Tension the halyard to remove wrinkles from the luff. If the luff is stretched, ease the halyard. In light air, an over-tight halyard hurts performance. As the wind builds, increase halyard tension to keep the luff firm. After sailing in strong winds, before rolling the sail ease the halyard to relieve luff tension.

Genoa (or Jib) Sheet: Trimming the sheet adds power by increasing angle of attack and by reducing twist. For close hauled trim, the middle leech of the jib should be parallel to the centerline of the boat. The foot of the jib should be a little rounder than the foot of the main, and the overall shape should match the shape of the main.

As you trim the jib, pointing will improve, with some sacrifice to speed. At the point where additional trim does not improve pointing the sheet is overtrimmed. Ease slightly – just a few inches – to optimize jib sheet trim.

Headstay Sag: The amount of sag in the headstay can be controlled with an adjustable backstay. A tighter headstay flattens the sail, while extra sag adds power. Leave the headstay loose in light air – just firm enough to keep from flopping in chop. Add tension as the wind builds. In strong breeze set the headstay as tight as you can.

Genoa Leads: The genoa leads change the genoa sheet position along the genoa tracks.

As an initial setting, adjust the lead so the sail luffs along its entire height as you pinch up above close hauled.

From this initial setting the lead position can be fine tuned to the conditions. You may want to move the lead forward to add shape to the foot of the sail, and to reduce twist to increase power. This lead-forward shape is preferred for light air or choppy conditions.

Deeper, more powerful

Flatter, less powerful

Headstay sag, controlled with an adjustable backstay, changes genoa depth and power. More sag adds depth and power, while a tighter headstay creates a flatter, less powerful shape.

Moving the leads aft reduces power for better performance in stronger winds. An aft lead position flattens the foot of the jib by pulling out on the clew. Think of it like the outhaul on the main. Moving the lead aft also increases twist, spilling power from the upper part of the sail. For heavy air sailing, we want the top of the sail to luff before the lower section.

Leech Cord: The leech cord does not control sail shape. It is intended to prevent leech flutter, which can quickly stretch the leech of a jib. Tension the leech cord just enough to stop flutter, and no more. If your sail has a foot cord, the same principle applies.

Jibs and Genoas

Large overlapping genoas are a nuisance. They are difficult to handle, hard to tack, easy to damage, and impossible to see around. A smaller jib is much easier to handle. On boats with large mainsails a genoa is an unnecessary burden. In all but the lightest conditions a working jib provides comparable performance (hey, we're not racing!), and in moderate to heavy air the jib performs better.

These photos show the impact of lead position on foot shape. On the left the lead is too far forward, creating a foot shape that is too round and a sail shape which will not perform well to windward.

On the right we see good foot shape and trim for optimum upwind performance achieved through proper lead position and sheet trim.

Initial lead position, with the genoa telltales breaking high just before they break low, and the jib shape matched to mainsail shape.

Moving the lead forward reduces twist and adds foot shape for extra power when sailing in light air or chop.

With the genoa leads aft the foot is stretched flat, and the top of the sail twists open. Power is reduced, which is desirable in heavy winds.

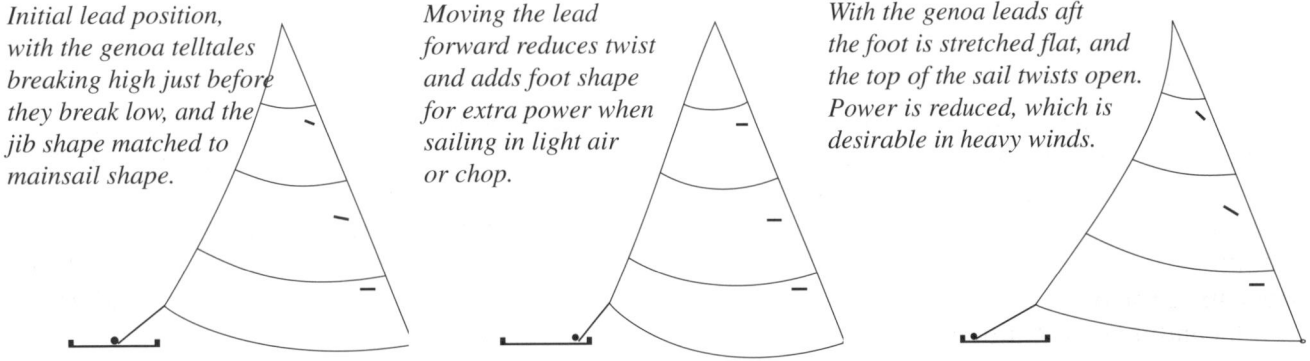

A working jib provides nearly all the performance of a genoa while being much easier to handle. You can even install a jib traveler which makes the jib self tending during tacks, like a mainsail.

Steering telltales: In most conditions the boat will sail best with the telltales streaming. When overpowered, feathering up will spill some power and allow the inside telltales to lift. If it is difficult to stay in the steering groove then the boat is out of trim – usually overtrimmed. Ease sheets an inch or two to add some twist and widen the steering groove.

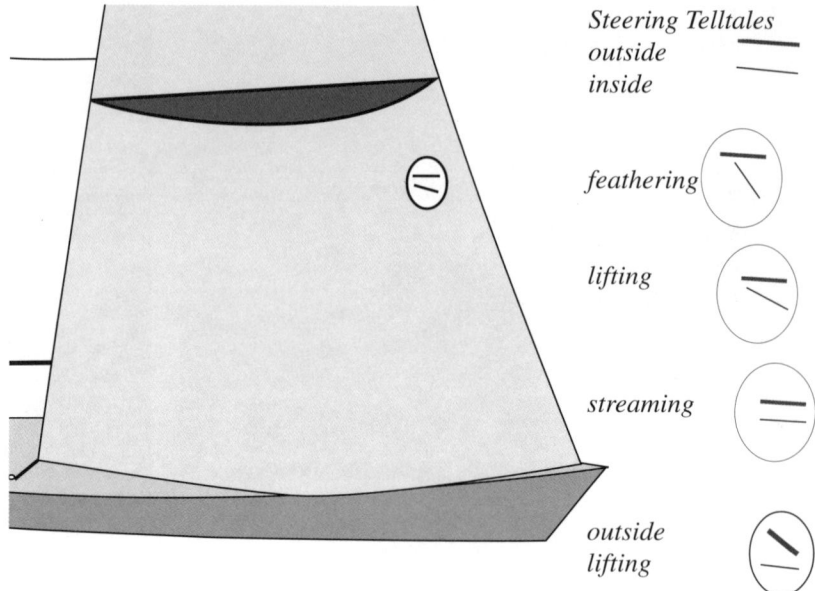

Steering Telltales
outside
inside

feathering

lifting

streaming

outside
lifting

5. Helming Upwind

There are a number of guidelines which can help you steer effectively upwind. There is also information you can glean from the feel of the helm to help you in trimming.

Steering Upwind

Depending on conditions, a good driver will reference some or all of the following when steering upwind:

• Jib Telltales

Are the jib telltales flowing, luffing, or stalled?

• Balance of the Helm

Is the helm balanced, with just a slight weather helm, or are you wrestling the helm to hold the boat on course?

• Angle of Heel

Is the boat sailing at a comfortable angle of heel? Is it overpowered and heeled too much, or underpowered and too upright?

• Boat Speed

Is the speed steady, rising, or falling?

• Apparent Wind Angle

Is the boat pointing well?

• "Feel" of the boat

Is the boat fully powered, or sluggish? Is the boat pitching in the waves, or punching through the chop?

Steering to Telltales

For starters try to steer the boat with the telltales flowing.

With the jib trimmed closehauled use the lower jib telltales as a steering guide. Steer so the telltales are streaming aft. Head up just short of the point when the inside telltales luff. The range of angle you can steer through with the telltales flowing - a range of a few degrees - is the steering groove. By paying attention to the feel of the boats power, and boat speed, you'll be able to tell how far up in the steering groove you can point while maintaining power and speed.

If you head up too high the inside telltales will luff, and soon thereafter the sail will start to luff as well. But even before that point you will start to lose power. Aim to steer as high as you can while maintaining full power.

If you fall off too far then the outside telltales will stall, and you will lose power and speed (and of course pointing as well, since you are falling off.)

When overpowered, head up slightly, and let the inside telltales dance. This narrower angle of attack reduces power.

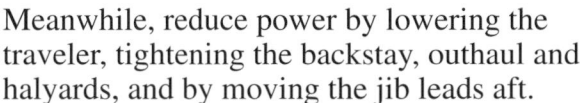
Sit to windward so you can see the jib, the horizon, and approaching waves, and feel the wind.

When sailing under auto pilot trim to balance the helm, and add extra twist to create a wider steering groove.

Meanwhile, reduce power by lowering the traveler, tightening the backstay, outhaul and halyards, and by moving the jib leads aft.

If you've got the telltales flowing, but the helm has no feel and the boat seems sluggish, then add power: Try traveler up, backstay, outhaul, and halyards eased, and jib leads forward. Also, bear off a couple of degrees.

In big chop, add power by falling off a couple of degrees until the outside telltales dance. Be careful not to fall off too far, or the outside telltales will stall and performance will suffer.

If it is difficult to hold the boat in the steering groove, then you may be over trimmed – ease the jib a couple of inches. With the telltales flowing, check your other guides.

Heel and Helm

Weather helm and angle of heel are key guides to upwind performance. If you must battle the helm, and the rail is in the water, then reduce power by adjusting trim and course. Flatten your sails, and head up to reduce angle of attack. In moderate to heavy winds you can sail with angle of heel as your primary guide. Steer to maintain consistent angle of heel and consistent power. Feather up (head up slightly) as you heel over in puffs, and foot off a few degrees to maintain full power in the lulls.

Battling Chop

If you are pitching excessively when steering through waves then bear off a few degrees to add more punch. If you are over powered when you bear off, then ease sheets a couple of inches to increase twist. The extra twist will spill some power, and give you more consistent power as you pitch, roll, yaw, and steer through the waves.

Chop can stop you dead in your tracks if you feather up in fresh breeze. Once again, rather than head up, add twist to spill power, while keeping the bow down for extra punch.

Trimming for an Auto Pilot

Before you turn the steering over to an auto pilot you may need to retrim to suit the system. Self steering works best with a well balanced boat and a wide steering groove. Set the boat up with slight weather helm, and trim your sails with a little extra twist to provide more steering latitude.

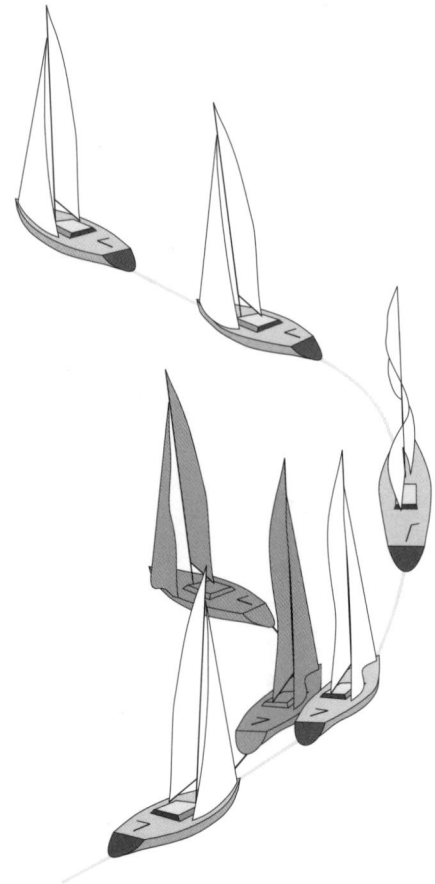

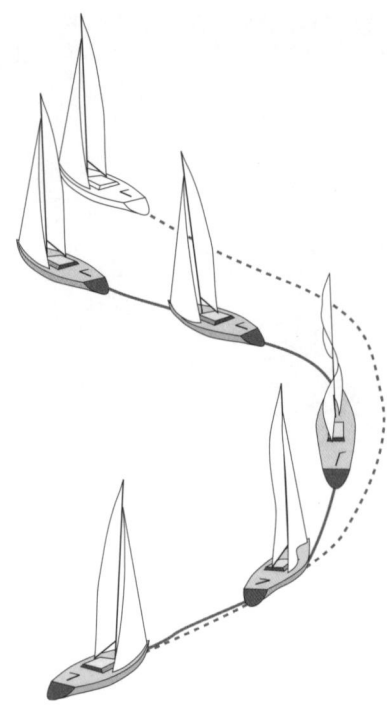

Tacking in waves requires a faster turn, as waves sap momentum. Tack as you climb the face of a wave, and push the bow around before the next wave hits.

Use a smooth turn to carry momentum through your tacks. Turning too fast throws off momentum, stopping the boat, and makes the jib difficult to trim.

Tacking

When Coming About think *not* about how quickly you can get the boat to the new tack. Think instead about carrying momentum onto the new tack. Too fast a turn – which is common – and you throw off your momentum. Too slow a turn, and you lose all your speed.

"Ready About"

Although we say "hard-a-lee," "soft-a-lee" might be more apt. Prepare to tack by checking that the working jib sheet is flaked and ready to run, and the slack is out of the lazy jib sheet, which is loaded with two full turns on the winch.

Steering Through the Tack

When tacking start with a slow smooth turn into the wind. This slow initial turn will help carry speed, and will also carry

you at nearly full speed toward your destination. As the sails luff turn more quickly to bring the bow through the wind. Once the bow crosses the wind ease back on your turn. Center the helm before you are down to the new closehauled course, as the boat's angular momentum will finish the turn for you. Position yourself well to windward (or to leeward in very light air) so you can see and steer to the jib as it is trimmed. Come out of the tack just a few degrees below your close hauled angle, and head up to course as the boat accelerates to full speed.

Handling the Jib

While you may want to clear excess wraps off the winch, be sure to keep the jib fully trimmed until it luffs half way across the foredeck. As the jib luffs, ease a full arms length of jib sheet to reduce load, then spin the rest of the sheet off the winch and make sure it runs.

On the trim side, take up slack, and as soon as the jib clew passes the mast, pull full armloads as fast as you can. When the sheet load is too great to pull, add wraps and grind

Release the jib as it luffs half way across the deck, and let the sheet run.

For quicker (and easier) trimming in heavy air, when coming out of a tack hold a high course until the jib is nearly fully trimmed.

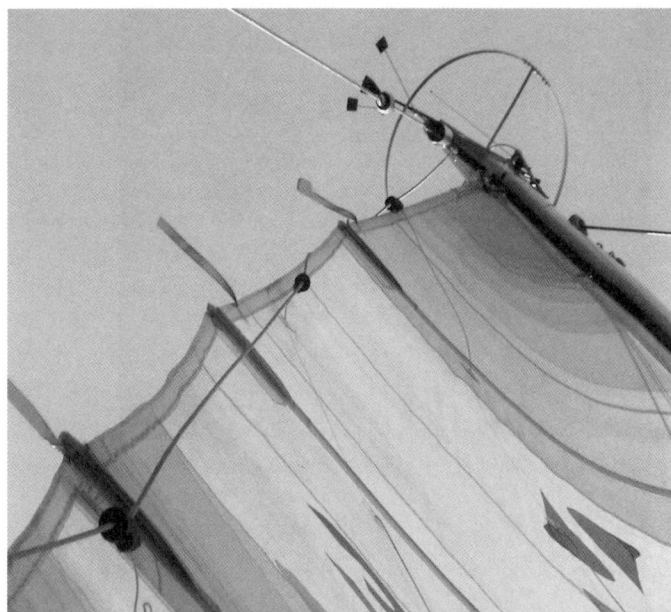

Main leech telltales flowing: Trim the mainsheet to get the upper leech parallel to the boom. For higher pointing trim until the telltales show an occasional stall. For extra speed and easier steering ease a few inches, and keep all the main leech telltales flowing.

the sail in the rest of the way. Stand up over the winch to grind, and use two hands.

For *best* performance it also helps to ease the mainsail a few inches for better acceleration coming out of the tack. Trim in again as you reach full speed.

Tacking in Waves

If time allows, look for a relatively smooth spot in which to tack, rather than tacking in the middle of a big set of waves. Use a quicker turn than in smooth water, as the waves will quickly sap the boat's momentum. Time your turn to start as you run up the face of a wave, and turn quickly enough to get the bow around so the next wave pushes you down onto the new tack.

Since big waves are generally accompanied by big wind, use the heavy air ideas which follow to guide you in finishing your tack.

Heavy Air

When coming out of a tack in heavy air you can make it easier to trim by slowing your turn. Hold a high angle so the jib doesn't fill until it is nearly fully trimmed. Turn just far enough to get the jib around the mast and shrouds, and then hold course with the jib luffing over the side deck. Don't bear off to fill the sail until it is nearly sheeted home.

6. Moderate Air Trim

In moderate winds of 8 to 10 knots we trim for full power. Here's how we set each source of power for each sail.

First the main:

Angle of Attack: Trim the main so the boom is on, or near, the centerline.

Twist: Trim the mainsheet so the telltale at the top batten is flowing most of the time, with an occasional stall.

Depth: Adjust sail depth for a slight weather helm and comfortable angle of heel.

For the jib:

Angle of Attack: This is primarily controlled from the helm, as we'll see.

Twist: Set the jib leads so that the jib shape matches the main, and the telltales break evenly. See below for further details.

Depth: If you can control headstay sag, tighten the headstay to flatten the jib and improve pointing, while maintaining enough power for a slight weather helm and comfortable angle of heel.

Sheet the jib hard to trim in against the rig, and set the leads for a balanced shape, parallel to the main.

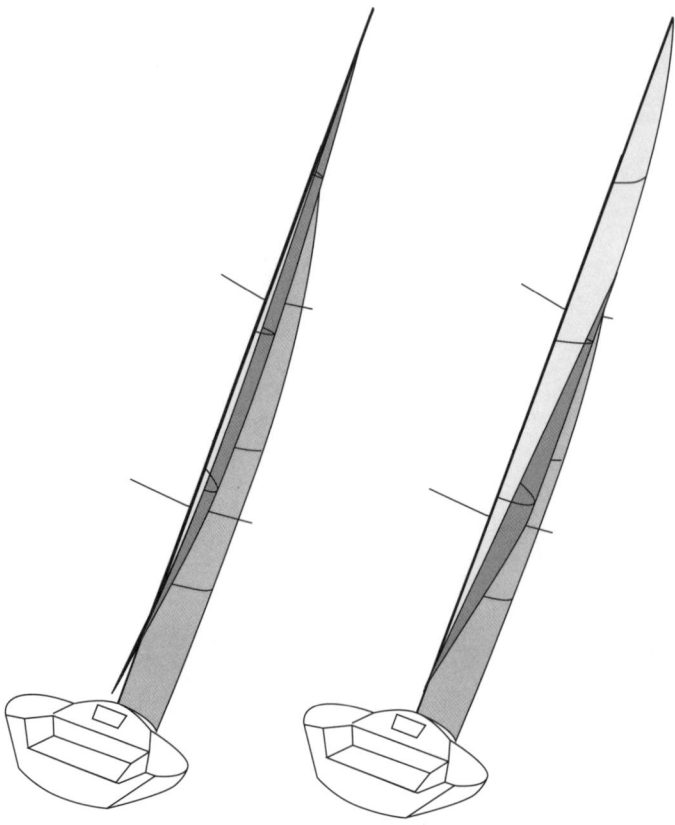

Moderate air trim fine tuned to sea state:
On the left:
In smooth water trim with little twist for high pointing.
On the right:
In chop, add depth and twist for better acceleration.

Jib Leads
Trimming to Telltales

Set the jib fairleads so the sail has a fair curve and even shape from top to bottom. The leech of the jib should match the shape of the main. When the sheet is trimmed, the jib telltales should break evenly from top to bottom. (As you pinch up above closed hauled, the upper telltales will luff *just before* the lower ones.)

NO, the telltales will not all break together. The upper telltales will luff first, and the break will spread down.

Balance of Power

The jib leads balance high and low shape in the sail. Our goal is to set the lead so the sail shape matches the wind from top to bottom. When the lead is set properly, the inside telltales will break smoothly, starting from top and moving down.

Moving the lead forward makes the sheet pull down more on the upper part of the sail, trimming in the top. This adds power for extra punch through waves.

Moving the lead aft will cause the sheet to pull back on the foot, like an outhaul, without trimming the upper part of the sail as much. This reduces power and allows harder trim for higher pointing in smooth water.

Steering to Telltales

Once the jib is trimmed we can use the lower jib telltales as a steering guide. Sail so the inside telltales are streaming aft. When overpowered, head up slightly, and let the inside telltales dance. This narrower angle of attack reduces power. In big chop, add power by falling off a couple of degrees until the outside telltales dance. Be careful not to fall off too far, or the outside telltales will stall and performance will suffer.

Sea State

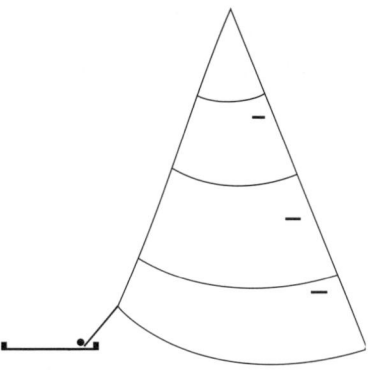

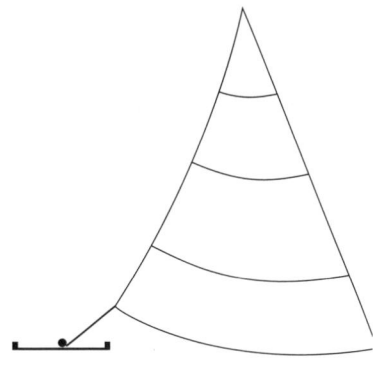

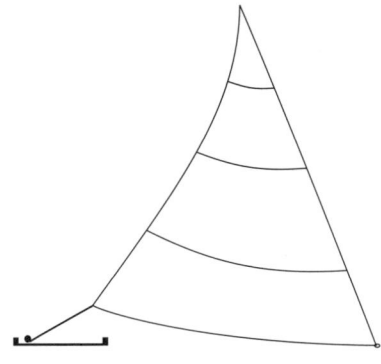

Trimming Telltales: Leads forward creates an even telltale break and a deep, powerful shape, best for light air or chop.

With leads middle the telltales break spreads from the top down, for moderate air sailing.

Moving the leads aft twists open the top of sail luff while pulling the foot flat, spilling power for heavy air sailing.

From our initial moderate air settings we can fine tune sail shape to match the sea state. The goal is to sustain full power, while adjusting the mix of power to match the sailing conditions. We trade one kind of power for another, while maintaining the same total power. Here's how:

Trim in Smooth Water

In smooth water, try a flatter sail shape, with less twist, for higher pointing. You can accomplish this with more mast bend, and a tighter headstay. Also, move the jib leads aft to flatten the jib, and then trim the jib sheet harder to take out the extra twist.

Setting the jib leads can be a source of confusion. Moving the leads aft flattens the foot and *adds* twist, but in smooth water we want a flat sail with *little* twist. We move the lead aft to get a flat shape, and then trim the sheet extra hard to take out the extra twist. Result: A flat sail with little twist.

With sails sheeted hard and trimmed flat, sail as high as you can (reducing angle of attack) while keeping full power. Don't pinch, or the boat will stand up and you'll lose power and speed.

Trim in Chop

For sailing in chop, add extra sail depth for extra punch through the waves, and add some twist for more consistent power as the boat pitches in the waves. Straighten the mast, sag the headstay, set the jib leads forward, ease the sheets a couple of inches to prevent stalling, and bear off a couple of degrees to increase angle of attack.

Once again, we must reconcile lead position with desired shape. We want a deep shape for power, and a *twisted* shape for a wide groove. Moving the lead forward adds depth, but *reduces twist*. We ease the sheet to *restore twist* while maintaining the depth and power necessary to punch through the chop.

The challenge in chop is having sufficient power to accelerate after each wave. Sailing with extra sail depth and a wider angle of attack keeps the power on, while the extra twist maintains flow as the boat pitches, and prevents excess heel.

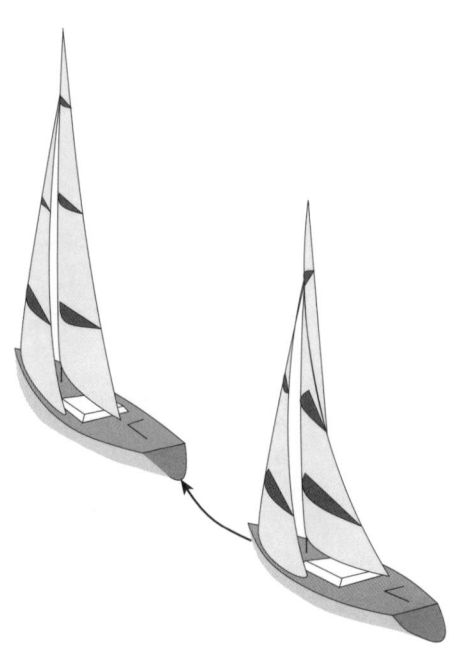

In light air bear off to a wide angle of attack, use deep sail shapes for power, and leave enough twist to encourage good flow. As speed builds trim the sheets to reduce twist, and head up for better pointing.

A point of sail that racers will never appreciate: Try motor sailing in light air to get where you need to go. The mainsail will help smooth out the ride, and it will also notify you when you've found enough breeze to shut down the noise maker and go sailing.

7. Light Air Trim

Building speed in light air can be a real challenge. From our moderate air baseline, here are some changes in trim to get full power for light air sailing.

Angle of Attack

Bear off - you can't point high in light air. Bear off a few degrees for power.

Sail Depth

Deep sail shapes are needed. Straighten the mast, and sag the headstay. Keep the headstay just tight enough so it doesn't flop around. Also move the jib leads forward, and ease the main outhaul. Finally, make sure you halyards are loose - leave a few wrinkles along the luff of each sail.

Twist

Leave enough twist to encourage easy air flow. If the sheets are overtrimmed then flow will stall. Light air requires plenty of twist - keep the telltales flowing.

Motor Sailing

In lighter breezes, especially with lingering chop, motor sailing can be a real boon. The extra speed not only improves the motion of the boat; it also creates more apparent wind for the main, which in turn provides more stability. Motor sail your way to fresher breeze. Hey, it's not a race.

One cautionary note: In no wind, you can overload and stretch your mainsail leech if you trim hard and let the sail slat / snap / flop as you motor through waves. Take the main down in next-to-no wind to prevent damage.

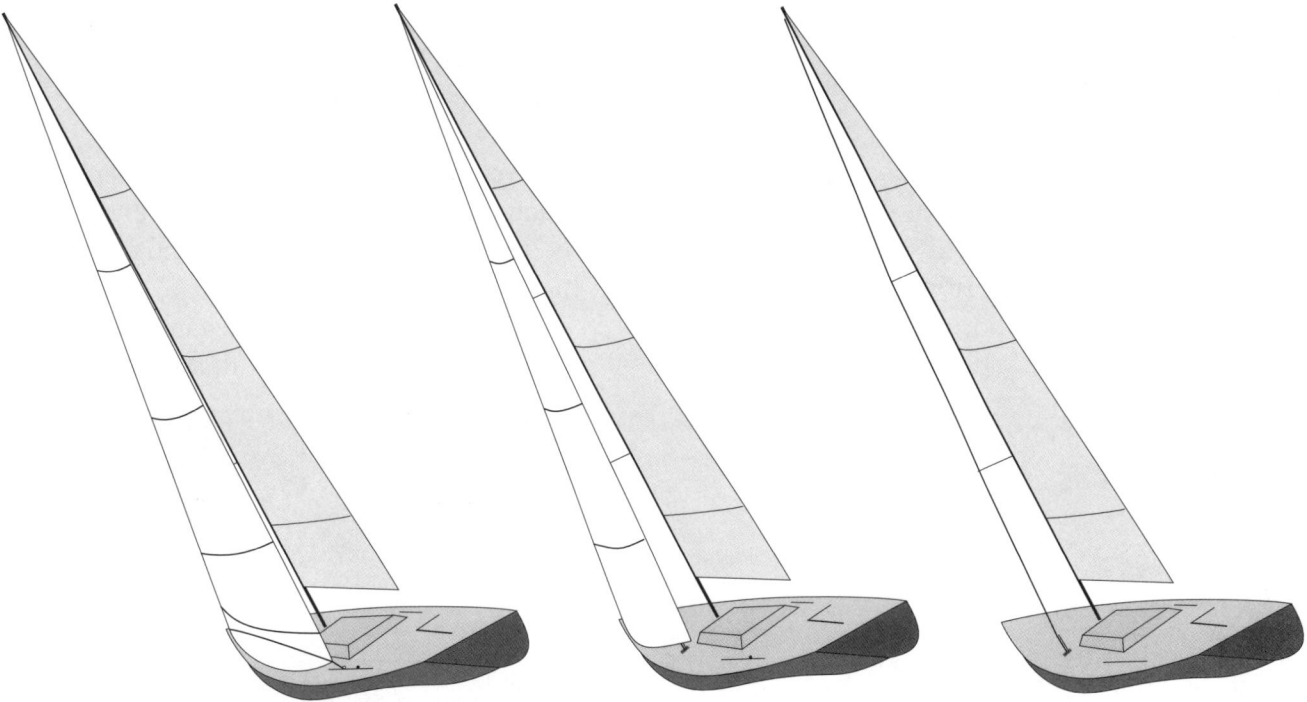

In heavy air upwind sailing you can roll up or change down to a smaller jib, reef the main, and if still overpowered, stow the jib entirely, and motor sail. Another option to consider: Pick an alternate destination.

8. Heavy Air Trim

Stronger winds call for changes in trim to reduce power. When the boat carries an uncomfortable amount of heel or strong weather helm it is time to depower the sail plan - but how?

The sequence for depowering runs something like this:

Reduce Sail Power
- Reduce power from sail depth
  - Bend the mast.
  - Tighten the outhaul.
  - Reduce headstay sag.
  - Move the jib leads aft.
- Reduce power from angle of attack
  - Feather the boat up (head up slightly)
  - Lower the traveler
- Reduce power from twist
  - Ease sheets a few inches

Reduce Sail Area
- Roll or change jibs
  - Roller-furl a few turns or ...
  - Change to a working jib

- Reef the main
- Stow the jib and motor sail

When properly trimmed, the boat will sail at a moderate angle of heel, with sufficient power to fight the prevailing seas.

Sea conditions play a big roll in trim. In flat seas very flat sail shapes or reduced sail area work well. In wavy conditions, the challenge is to keep enough power to punch through the waves without being overpowered.

Overpowered!

You've done all you can to reduce sail power with the sails you've got up, and you are still overpowered.

Now what?

It is time to reduce sail area, but in what order? Generally it is best to reduce jib size first, then reef the main, and finally stow the jib and motor sail. See *Chapter IV - Heavy Weather Sailing*, for details.

9. Performance Problems and Trim Solutions

Poor speed

If boatspeed seems poor then you may need to add power. Try deeper sail shapes, and bear off a couple degrees.

You can also be slow from being over powered, with too much heel and weather helm. Reduce power and balance the helm to restore speed.

A fouled bottom can also have a big impact on speed.

Poor pointing

If your boatspeed is good, but the boat is pointing poorly, then try sheeting harder. Trim the jib until you get a hint of backwinding in the main, and trim the main to the verge of stalling the top leech telltale.

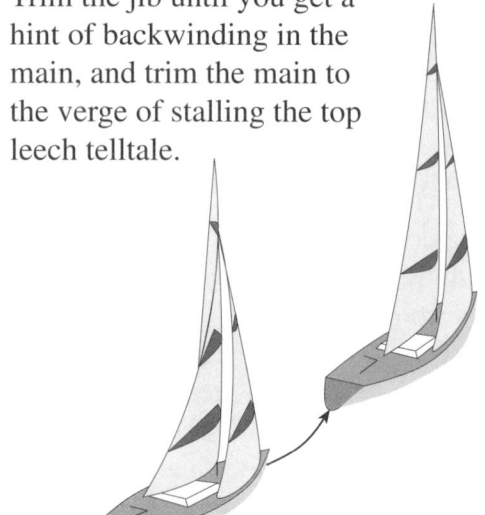

Too much weather helm

Reduce heel to reduce weather helm. Try flatter sails, and more twist. Also, feather up to reduce angle of attack.

Weather helm may also mean there is too much power in the main, and not enough in the jib. Add power to the jib, and ease the main.

If weather helm is a constant problem, then try tuning your rig with less rake.

Control weather helm and heel by reducing power. Start by flattening sails (as above). If that is not sufficient, roll or change to a smaller jib (as below).

Telltales too erratic

If the jib telltales are too erratic, with both the inside and outside telltales dancing, then create a wider entry angle and more forgiving steering groove by easing the sheet and/or tightening the halyard.

With the draft properly positioned, as shown here, the telltales will settle down.

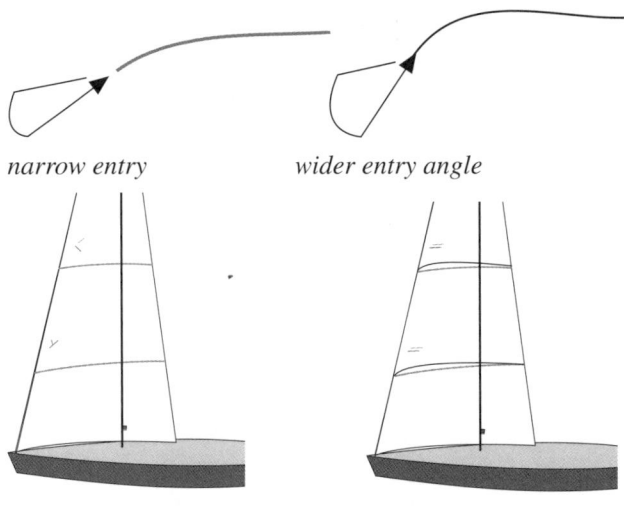

narrow entry   *wider entry angle*

Too little weather helm

If you have no feel in the helm, then add power to increase weather helm. Try deeper sails, and bear off a few degrees. Also try moving weight to leeward to increase heel.

In light air a lifeless helm may be a sign of being over trimmed. Ease the mainsheet and jib sheet, and bear off to add power and speed.

If a lack of helm is a consistent problem, then consider retuning your rig with more rake.

Pounding and pitching in waves

Sailing into waves requires power - power to punch through the waves. If the boat is pounding, then foot (fall off) slightly and add twist to keep from being overpowered. Also, moving weight (like anchors) off the bow and out of the forepeak can reduce pitching markedly.

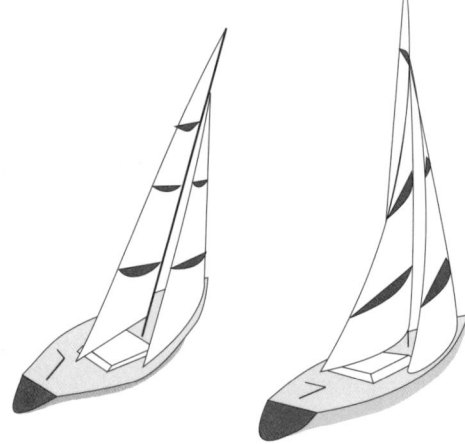

Overworked Auto Pilot

If your auto pilot is constantly searching, and can't seem to settle on course, and your boat is upright and underpowered one moment, and overpowered the next, then add twist. Twist creates a more gradual onset and release of power.

When your auto pilot is working too hard, retrim for better balance, and add twist to smooth out the transition from over powered to under powered.

Also, if you have the option, set your auto pilot to sail to the apparent wind angle when sailing upwind, rather than compass course.

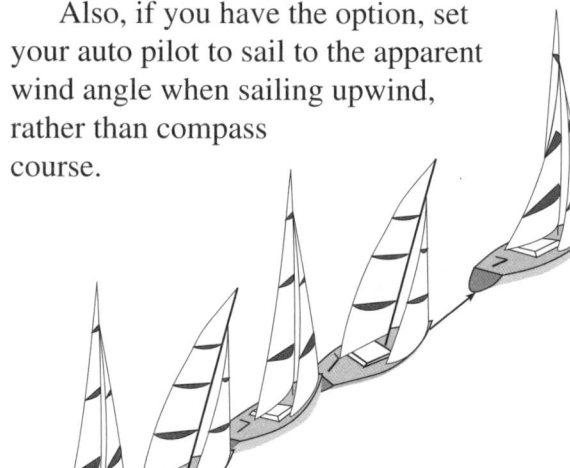

Extra twist, as in the lower photo, reduces pitching in waves and improves auto pilot performance.

Overpowered

A small jib and reefed main provide a balanced rig for best performance in heavy air conditions. See Chapter IV for more on *Heavy Weather Sailing*.

Chapter III –
Downwind Sail Trim

1. **Reaching Trim**
2. **Running Trim**
3. **Gennakers and Spinnakers**

Downwind Sail Trim

As the genoa sheet is eased onto a jib reach, the genoa lead should follow the sail outboard and forward.
Left: Close hauled trim.
Middle: If the lead is not moved as the jib is eased then the foot will be too round, and the leech will spill.
Right: A proper lead position will keep the top of the sail from spilling open and move the genoa foot outboard, allowing the main to be eased properly.

The boom vang controls mainsail twist on a reach.
Left: With the vang eased, the boom rises, and the sail spills power.
Right: With the boom vang trimmed tight the leech remains in trim. Power and speed are maintained.

1. Reaching Trim

Close Reaching

As you bear away from close hauled to a close reach the forces on the sails rotate forward, speed jumps, and heeling forces are reduced. To make the most of the wider wind angle, retrim the sails for the new course.

Ease the jib, and for best performance move the lead outboard and forward, chasing the clew of the sail with the lead. Keep the halyard firm to hold the draft forward and prevent the back of the sail from becoming too round.

If the lead is not moved as the sheet is eased, then the top of the sail will twist open, spilling power, and the bottom of the sail will hook in toward the boat, creating excess drag.

With the jib trimmed outboard, ease the main. Keep the vang tight, and ease the mainsheet or lower the traveler. As the main goes out, heeling forces decrease, and the boat accelerates.

As the boom goes out, the vang is critical to control twist on a reach. As an initial setting, take the slack out of the vang while trimmed close hauled. Then, as you turn to a reach and ease the mainsheet, the vang will prevent the boom from rising and the leech of the main from spilling. When overpowered on a reach, easing the vang will spill power, reduce heel, and balance the helm.

Ease the mainsheet to keep the leech telltales flow. If the sail luffs, then trim in. Also, ease the outhaul and backstay slightly to add power to the main.

On a heavy air "blast reach" keep the main flat, and ease it to half luffing, if necessary, to control heeling forces and weather helm. If still overpowered, then reef. On a close reach, a reefed main and big jib can be an effective sail combination.

*Close Reaching Trim:
Ease the sheets, and chase
the jib clew by moving the
jib lead forward. Ease the
mainsheet, and control
mainsail twist with the
vang.
On the left we see too much
jib twist, and not enough
main twist.
On the right we see
excellent reaching trim.
The sails have matched
shapes with moderate twist.*

Beam to Broad Reaching

As we bear off further, the boat stands upright. Ease sails. On a beam to broad reach the top of the jib will spill open. Trim to keep the middle of the sail working.

Ease the main until it luffs. The main should go way out - out against the rig if necessary. If the sail doesn't luff, let it rest against the rig. It won't hurt the sail or the rig. Keep the boom vang firm enough to hold the top batten of the main parallel to the boom.

Reaching in a beam or following sea can be treacherous, as the boat lurches around. A preventer (a line that holds the boom out) should be rigged to control the boom, and prevent the boom from swinging wildly. Auto pilots are at their worst in these conditions, as are human steerers. A spinnaker can add extra power, speed, and stability. Short of that, motor sailing is an option – or consider a new destination...

Don't forget to ease the main. Snug the vang, and ease the mainsheet just shy of luffing.

Sailing Wing and Wing with the genoa sheet running through the pole end. Top left: The pole is held in place by a topping lift, (after) guy, and foreguy.

Top right: Poling the jib out allows you to sail above dead downwind, giving a broader choice of courses, and reducing the chance of an accidental jibe.

2. Running Wing and Wing

Running dead before the wind under jib and main requires careful steering to avoid an accidental jibe. In light air, it often pays to reach up, and sail with the jib in normal position, as the extra speed will make up for the extra distance.

Running wing and wing, a whisker pole will help the genoa fly much more effectively. The pole should be set with topping lift, afterguy, and foreguy to hold the pole in place. With the pole trimmed to position, the genoa sheet is run and trimmed through the end of the pole.

In heavy air there are often following seas, which can make steering a challenge. A poled out jib allows sailing above a dead downwind course, providing a wider steering lane, and reducing the chances of an accidental jibe.

Opposite the jib, rig a preventer for the mainsail. Run the preventer from the end of the boom to a block well forward on the rail, and then back to the cockpit, near the mainsheet cleat. Every time you ease the mainsheet, tighten the preventer. It doesn't have to be bar-taut – just tight enough to prevent the boom from jumping around. If it's led this way, the preventer can easily be cast off as necessary.

To jibe when sailing wing and wing a slow turn downwind is all that is needed before jibing the main. And don't forget to release the preventer.

See Chapter IV for more on heavy air sailing and preventers.

Jibing

When our destination is downwind on the opposite tack then a jibe is called for.

A jibe can be considered in three parts:

1 - Starting from a broad reach initiate the jibe with the command **"Prepare to jibe." Release the preventer** and **turn slowly downwind**.

2 - When the wind is dead astern **the jib will jibe itself**. This is the signal to hold a **steady course**, pull the mainsail amidships, and then ease it all the way out on the new tack with the hail *"Jibe Ho."*

3 - After the sails are across, continue the turn to your new course.

More Details

The Jib is the Signal

The jib is the clue to a successful jibe. Once the jib comes across on its own, trim the new jib sheet and hold a steady course while bringing the mainsail across by hand.

Use a slow turn

Keep control of the mainsail so the boom will not fly across. A slow turn allows time to trim and control the main.

Jibing

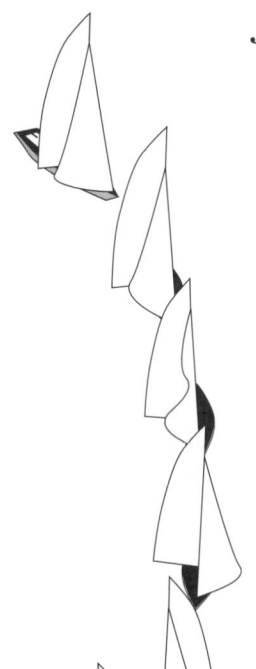

From a broad reach hail "prepare to jibe" and …

…turn slowly downwind.

As the jib collapses slow your turn …

and as the jib jibes itself hold your course…

…and jibe the main.

Trim your sails on the new tack and complete the jibe by turning to your new course

Ease the main
As the main jibes ease the mainsheet all the way out - just let it run.

Watch your course
In stronger breeze, as the main jibes it will load the helm, and try to turn the boat. Watch your course, and correct the helm to keep the boat from rounding up.

In light air as the jib jibes grab all the parts of the mainsheet and fling the main across to the new jibe.

Don't shy from jibing.
In all but the breeziest conditions, a well executed jibe is a safe and effective way to change tacks.

Prevent an Accidental Jibe
If the jib jibes itself unexpectedly, it is a signal that the main may soon follow. To prevent an uncontrolled jibe, turn up immediately. Straighten out once the jib returns to its normal position. As a precaution always *keep your head down* when you see the jib cross the boat, and use a preventer to secure the boom.

The jib signals the jibe: As the jib collapses and jibes itself it signals the time to jibe the main. Use a slow turn to allow plenty of time to handle the main.

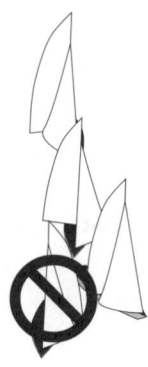

If the jib jibes unexpectedly, head up to prevent an accidental jibe.

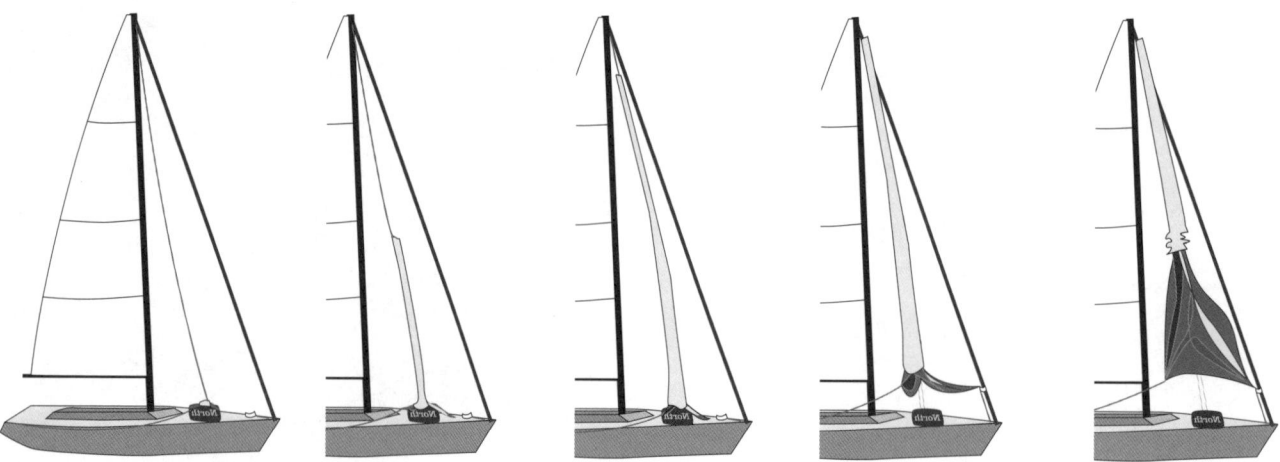

To set a Gennaker, first hoist the Snuffer to full hoist, pull the tack to the forestay, pull up the Snuffer and trim the sheet. To douse turn to a broad reach, ease the sheet, Snuff the sail and lower it.... (Figure runs across spread)

3. Gennakers and Spinnakers

Spinnakers provide an enormous performance boost in light to moderate air downwind sailing – and well they should, considering the trouble they can cause! We'll take a look here at how to handle and trim both cruising spinnakers – also called *Gennakers*, which fly without a pole, and conventional spinnakers with poles.

Gennakers

Gennaker Sets

A cruising spinnaker or Gennaker is set with a tack line from the bow, a halyard, and a sheet lead to the quarter. For shorthanded cruising, a spinnaker Snuffer is recommended. The sail is hoisted in a protective sock, and once up, the Snuffer line is pulled to retract the sock and free the sail. To prevent twisting, the tack should be pulled to the forestay prior to the hoist.

Depending on the luff length of the Gennaker, it may also be advantageous to rig a tack strap. Rigged around the rolled jib, the tack strap prevents the tack from wandering. If your sail has long luff and low tack – just above the bow pulpit – then a tack strap is not required. For a sail with a short luff and high tack, a tack strap adds control.

Gennaker set up: Rig the tack to the tack line forward. A tack collar, as shown, is only necessary if the Gennaker flies well above the pulpit. Rig the sheets to the clew, and lead them outboard and well aft. The halyard attaches to the top of the Snuffer.

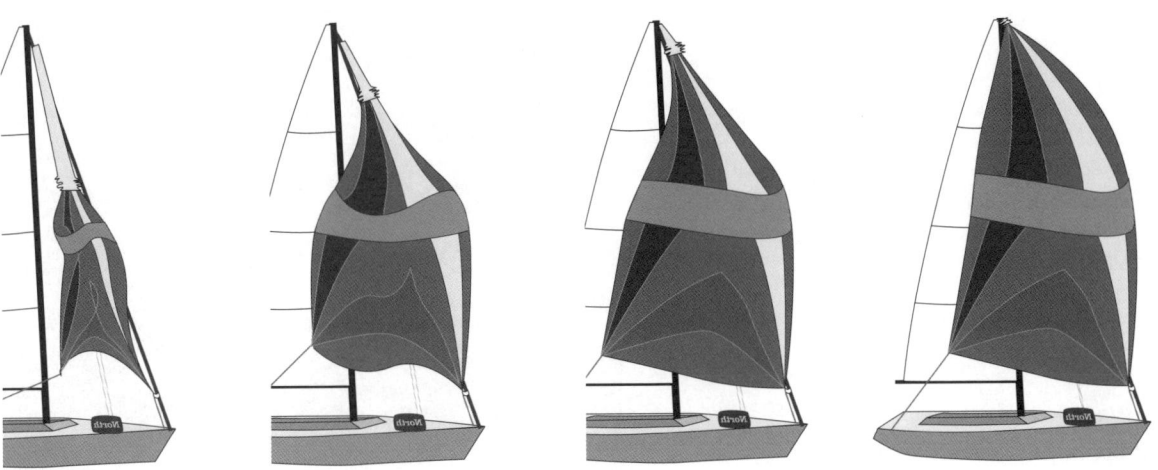

To hoist turn to a broad reach and raise the Snuffer. At full hoist pull the snuffer up and trim the gennaker sheet. To prevent their fouling, put the Snuffer control lines into the sleeve during the set up and hoist. Make sure the control line sleeve runs straight up, and does not "barber pole" around the Snuffer.

These photos show a gennaker jibe. Note how the new sheet is trimmed and the sailed pulled to the new leeward side before the turn is completed and the mainsail is jibed. This boat was rigged with the sheets inside the luff.

Gennaker Jibes

There are a couple of ways to jibe a gennaker. One method is to snuff the spinnaker, re-lead the spinnaker sheet to the new leeward side, and redeploy after the boat and mainsail have been jibed. The Snuffer and Gennaker go around the outside of the forestay on a jibe. The disadvantage of this jibing technique is that you must go forward to pass the snuffer around the headstay. (Only if the Gennaker is flown from a jib halyard, rigged under the forestay, would you take the Gennaker inside the forestay on a jibe.)

The Gennaker can also be jibed flying. Square down to a very broad reach, tension the windward gennaker sheet, and haul hard as you release the working sheet. Pull and pull and pull until the sail collapses, inverts, and starts to trim back on the new jibe; then finish the turn and jibe the main. The trick is to have the turn follow the trim – trim the gennaker most of the way through the jibe before jibing the boat.

If the turn is too fast (or trim too slow), and the boat is jibed before the sail is trimmed, then the spinnaker can blow through behind the forestay, or it can wrap on itself, or the forestay.

The Gennaker sheets can be lead inside or outside the Gennaker luff. There are advantages to each set up, and both work....

To Snuff the Gennaker turn to a broad reach to blanket the gennaker behind the main, ease the sheet, and pull down the Snuffer line. Once the sail is snuffed you can lower it into its bag or down the forward hatch. When working on the foredeck, sit down so you can't fall down.

Gennaker Takedowns

To take the sail down, turn to a very broad reach to hide the Gennaker behind the mainsail, ease the sheet until the sail carries a big curl, and pull the Snuffer down over the sail. Once the sail is snuffed you lower the halyard and stuff the sail into its bag on deck, or pass it down the forward hatch to be bagged below.

One important detail when working on the foredeck: Sit down while you pull the sail down. If you sit, you can't fall. When gathering a sail on a rolling boat, lurching around, stepping on slippery sail cloth it is easy to fall down – or overboard! Sit Down.

The Snuffer in action. To douse, turn to a broad reach, ease the sheet, pull down the Snuffer, and lower the halyard. This sailor should sit down!

The Gennaker flies best on a reach. Ease the sheet until the luff curls, and then trim to take out the curl. Unlike a conventional spinnaker, a gennaker need not be flown carrying a curl. When the luff length puts the spinnaker tack just above the pulpit, as it is here, a tack strap is not needed.

Gennaker Trim

Cruising spinnakers are remarkable sails for their ability to change shape to match the course and wind angle. By trimming both the tack line and spinnaker sheet, we can transform the sail from a genoa, to a spinnaker, and back again.

Gennaker Sheet

The rule for sheet trim is pretty straightforward. Ease the spinnaker until the luff curls, then trim to remove the curl. For best performance, a symmetric spinnaker must be carry a curl. In contrast, a gennaker will deliver top performance without trimming to the verge of a luff. Ease to a curl to make sure the sail is not overtrimmed, and then trim to remove the curl.

For optimum performance, play the spinnaker constantly. Me? I check it between naps.

Tack Line

It is changes to trim of the tack line which allow us to change the shape and performance of the sail.

On beam reaches, keep the tack line snugged down and the luff of the spinnaker pulled firm. The Gennaker assumes the shape of a great reaching genoa.

On a broad reach the tack line is eased, allowing the tack to lift. Easing the luff this way lets the Gennaker roll out from behind the mainsail and assume a more powerful spinnaker-like shape.

On a broad reach in fresh breeze, ease the sheet to allow the gennaker to rotate out from behind the mainsail. Active steering is required to maintain course in these following seas. These are not conditions for an auto pilot.

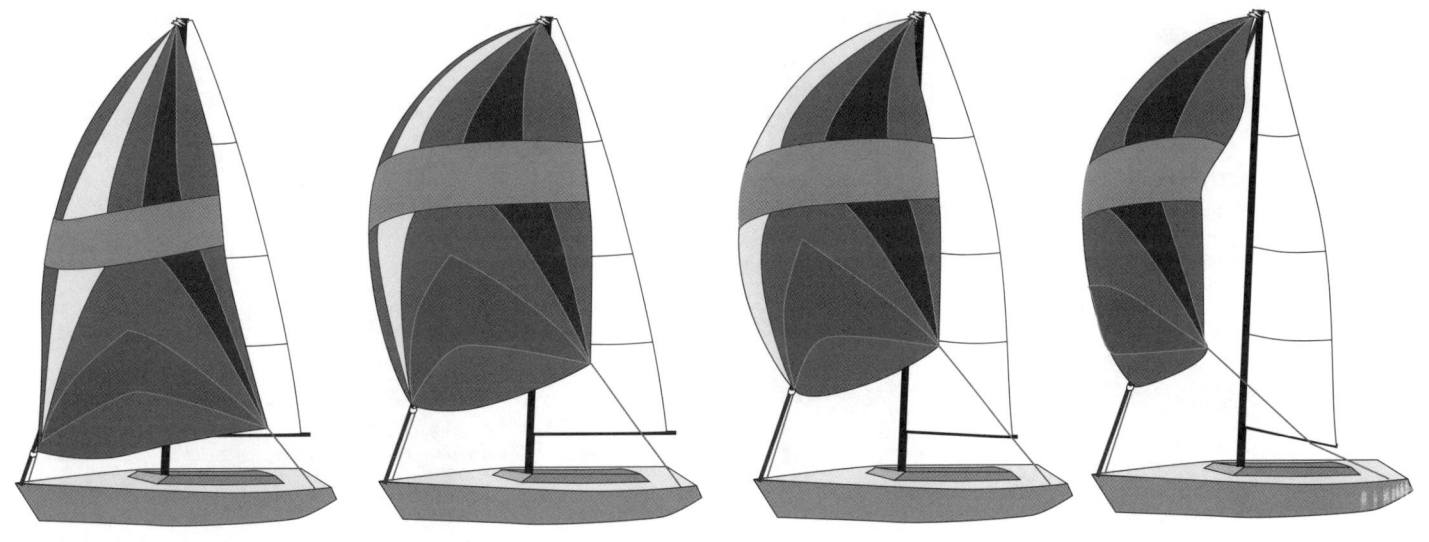

Two Views:
The Cruising Spinnaker morphs from a Gennaker shape to a spinnaker shape and rolls out to weather as the course changes from a beam to broad reach, and the sheet and tack line are eased.

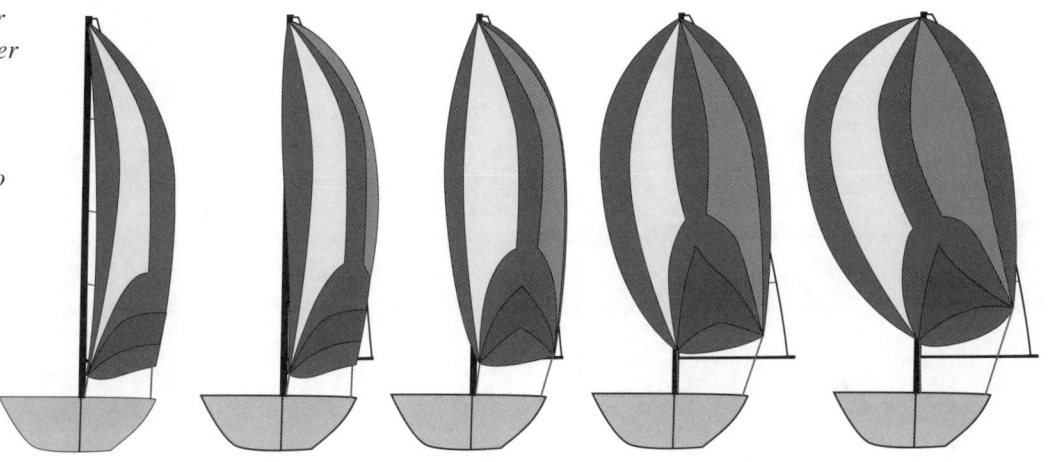

Tack height is used to control Gennaker shape. On a beam reach, at left, the tack is set at moderate height. On a broad reach the tack line is eased, creating a more powerful - spinnaker line shape, at center. On a close reach the tack is pulled down, creating a better reaching shape. The tack collar shown is only needed on Gennakers flying well above the pulpit. A longer luffed Gennaker, with a lower tack, would not need a collar.

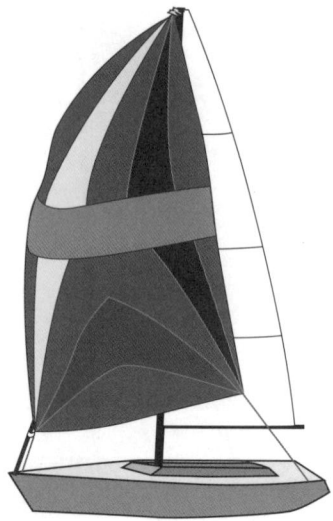
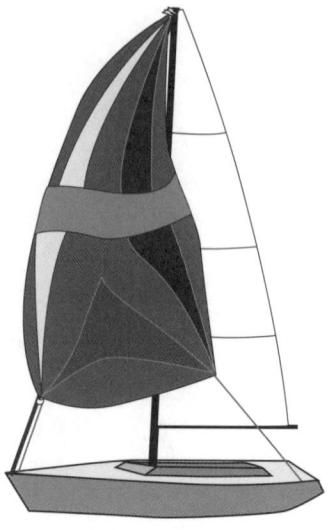

Trim to a curl   Over trimmed, or "Starved"

Gennaker Trim: Ease to curl the luff, as on the left, and trim just to take out the curl. If the Gennaker is overtrimmed, the sail will be "starved." The entire sail will sag, and the leech will cave in, as on the right.

The Gennaker on a beam reach, with the tack snugged down, and on a broad reach (lower photo),with the tack eased and luff rotated out from behind the mainsail.

<< On a beam reach, the tack is trimmed down to hold the Gennaker in a genoa shape.
>> On a broad reach, the tack line and sheet are eased and the sail rolls out to windward, assuming a spinnaker-like shape.

This cruising cat shows Gennaker trim on a beam reach (left), and a broad reach (center) with the sheet eased, and the tack rolled out to weather. On the right the tack is hauled to windward, which helps get the Gennaker out from behind the main.

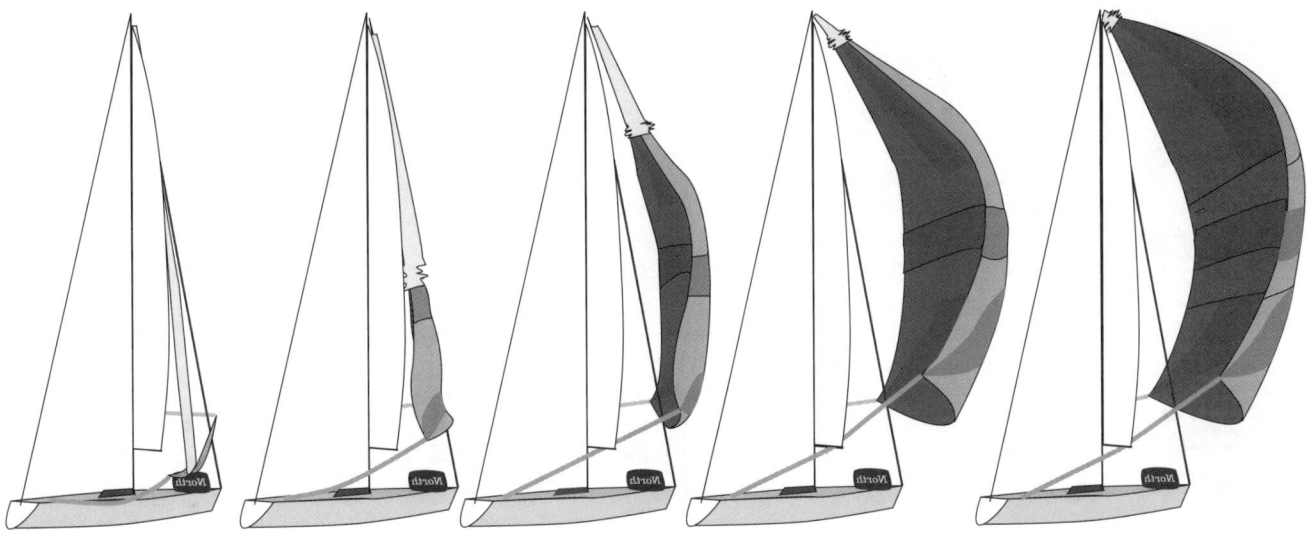

Hoisting with a Snuffer. Set the pole, and rig the sheet, guy, and halyard. Pull the tack to the pole, and hoist the sail in the Snuffer. Pull up the Snuffer, trim the sheet, and square the pole back. To douse, reverse the process.

Spinnaker Handling

Conventional spinnakers with spinnaker poles offer more control over spinnaker trim and sail shape - but it comes at the price of additional complexity.

For shorthanded sailing a Snuffer is recommended. Setting the spinnaker requires rigging the pole with a topping lift - to hold the pole up - and a foreguy - to pull the pole forward. Some skippers rig an afterguy directly to the pole to hold the pole back. Others allow the spinnaker guy to handle this function. You can sail safely and successfully either way. The windward spinnaker sheet - called the guy - is rigged to run through the end of the pole.

As you prepare to hoist, trim the guy to pull the tack of the spinnaker to the end of the pole, and then hoist. As you reach full hoist, take slack out of the sheet to prevent twists as you pull up the Snuffer, and then raise the Snuffer. When the Snuffer tops up, tie off the Snuffer lines loosely near the mast base, and trim the spinnaker sheet.

Set the pole perpendicular to the apparent wind, set the pole height to get the clews even, ease the sheet to a curl and trim, and then work on refinements of sail shape, as detailed below.

Short handed jibes are a challenge. The easiest technique is to snuff the sail, drop the pole, and pull the snuffed spinnaker around the bow with the sheets. Jibe the main, reset the pole, and then redeploy the spinnaker by pulling up the Snuffer.

To douse the spinnaker with the Snuffer reverse the hoist sequence: turn to a broad reach, ease the sheet, pull down the Snuffer, ease the guy, and lower the halyard.

In stronger winds there is an alternative technique which will better hide the spinnaker, and make the snuffing process easier: Once again, the process starts on a broad reach. Rather than ease the sheet, put a loop of line around the sheet, and use this "choker" to pull the clew of the spinnaker in close behind the mast. Ease the guy to luff the spinnaker, and pull down on the Snuffer. By pulling the leech of the spinnaker in close behind the main this technique assures that the spinnaker will be fully controlled and blanketed behind the main. As always, sit down on deck while pulling down the Snuffer.

To douse the spinnaker in strong winds turn to a broad reach and choke down the sheet behind the main. Ease the guy to luff the spinnaker and pull the Snuffer down over the sail.

By choking the spinnaker sheet down behind the mast the spinnaker is blanketed behind the main (2 photos below). The photo sequence at left shows the Snuffer in the shadow of the main, tracing directly down the leech of the spinnaker.
Note that the crew is sitting on deck.

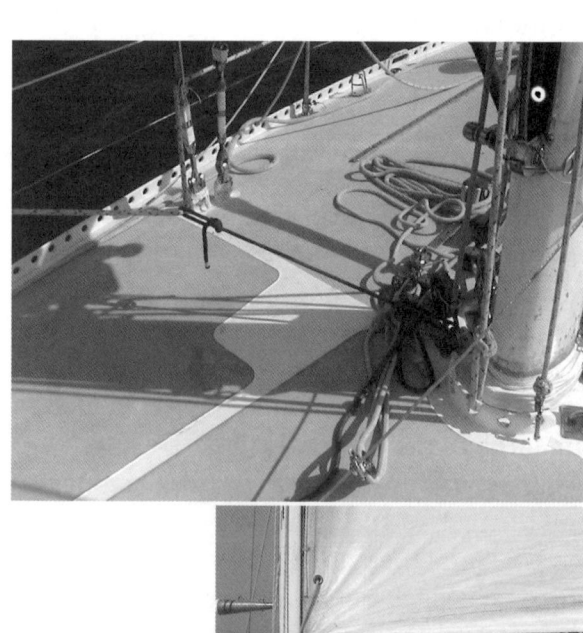

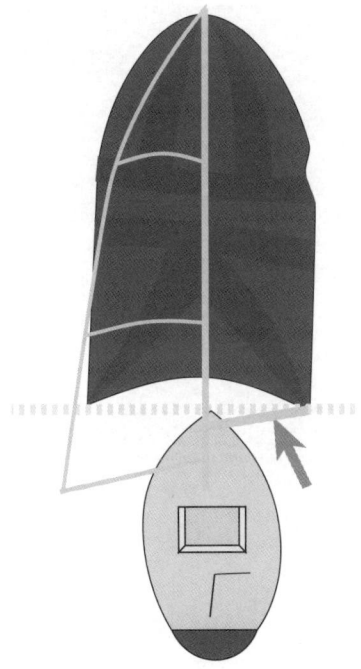

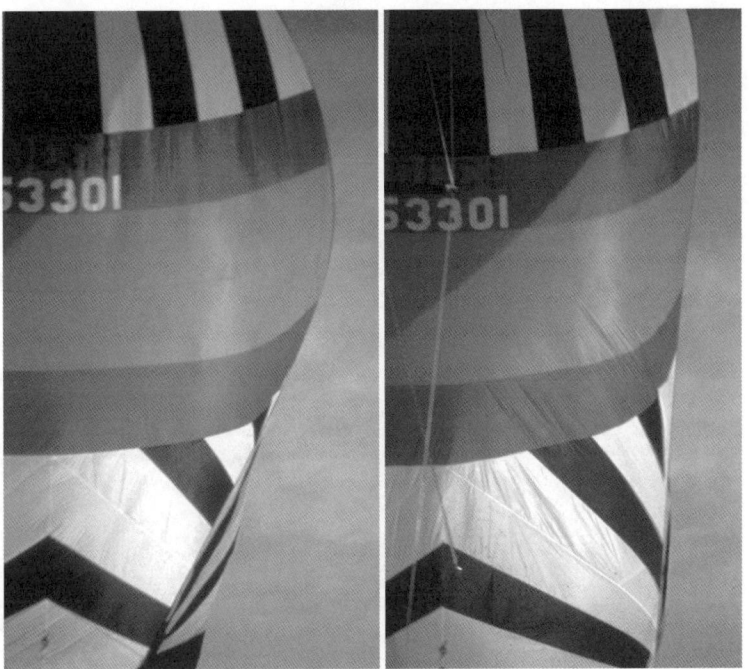

Initial spinnaker trim: Pole perpendicular to the wind, clews even, and play the sheet.

Refinements in spinnaker trim: On the left, the foot is too round, and the tack is inside the shoulder. By pulling the pole aft, as on the right, we get a better foot shape and a vertical luff.

On a beam, reach the tack should be lower than the clew, as in the upper figure. On a broad reach in fresh breeze, choke down the pole and clew to maintain control, as shown in the lower figure, and in photos on the next page..

Spinnaker Trim

There are three initial settings for spinnaker trim, and refinements from there which can improve performance. For starters:

• Trim the guy to set the pole perpendicular to the wind.

• Set the pole height so the clews are even.

• Ease the sheet to curl the luff, and trim to just take out the curl.

From these initial settings we can fine tune trim in several ways.

Generally, the guy can be trimmed aft further than square to the wind. Trim the guy to get the luff to run vertically from the shoulder of the spinnaker down to the tack, and to get the shape across the foot to match the mid shape.

If the shoulder is rolled out, and the foot is too round, then pull the pole aft. If the foot is stretched flat, and the tack is poking out to windward, then ease the guy and let the pole forward.

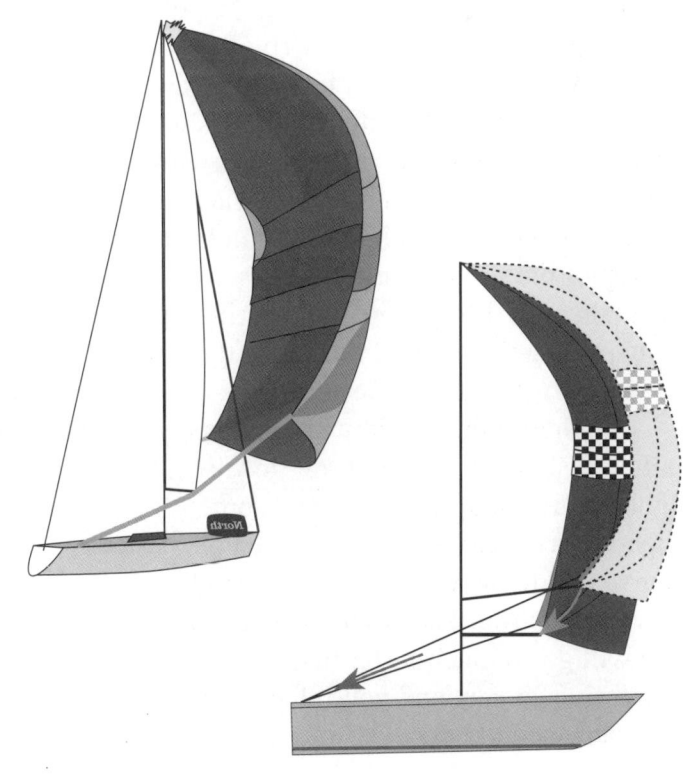

This spinnaker shows nice shape, with a vertical profile, and a foot shape matching the mid-shape.

The foot of this spinnaker is too round. The pole is also too high. Trimming back and down on the pole would improve the shape, as shown.

The sail below is too far from the boat. Even though the clews are even, the pole is too high and forward. Note how the top of the sail is blown out horizontally. The drawn shape is achieved by pulling down and back on the pole while also trimming the sheet.

Pole height can also be fine tuned. On a close or beam reach, try lowering the pole so the tack is lower than the clew (by a foot or more). This pulls the draft forward, and opens the leech, for a faster reaching shape. On a broad reach fly the pole higher, with the clews even. In fresh breeze beware flying both clews too high, which in turn lets the top of the sail flow out flat. Try to maintain a vertical profile by lowering the pole, pulling the pole aft further, and by choking down the sheet lead as well.

In lighter going the key to success is sailing fast angles. Do not sail downwind in light air - reach up until you feel some apparent wind from the side. Often the best course is fully 40 degrees above straight downwind. You'll sail extra distance, but you will go so much faster that you'll arrive downwind sooner, and you'll have a more enjoyable sail to boot!

Chapter IV –
Heavy Weather Sailing

1. **What is Heavy Weather?**
2. **Heavy Weather Forehandedness**
3. **Heavy Weather Sailing**
4. **Squall!**
5. **Storms**
6. **Alternate Storm Strategy**
7. **Misery and Danger**

Heavy Weather Sailing

The Beaufort Scale of Wind Forces

Force	Wind Speed - Knots -	Description
0	0	**Calm** - Smooth, like a mirror.
1	1-3	**Light air** - Small ripples, scales.
2	4-6	**Light breeze** - Wavelets, no crests.
3	7-10	**Gentle breeze** - Wavelets, some crests.
4	11-16	**Moderate breeze** - Small waves with some whitecaps.
5	17-21	**Fresh breeze -** Moderate waves with many whitecaps and some spray.
6	22-27	**Strong breeze** - Large waves, extensive whitecaps and spray.
7	28-33	**Near gale** - Heaps of waves, with some breakers. Foam blown downwind in streaks.
8	34-40	**Gale** - Moderate to high waves with heavy spray and foam in strong streaks.
9	41-47	**Strong Gale** - High waves, dense foam streaks, and rolling crest. Spray reduces visibility.
10	48-55	**Storm** - Very high waves with overhanging crests. The sea looks white. Visibility greatly reduced. Waves tumble with force.
11	56-63	**Violent storm** - Exceptionally high waves, froth and foam.
12	64-71	**Hurricane -** The air filled with foam and spray, and the sea is entirely white.

1. What is Heavy Weather?

Heavy weather is not an absolute wind strength. It's when the wind and waves are such that you risk losing control of your boat. This may happen to a novice crew in a small boat in a moderate wind of only 12 knots. But an experienced crew in a big cruiser may not have control problems until they're in a 35 knot gale.

Your course and speed are factors. In a moderate 12 knot true wind, a boat sailing closehauled at a speed of 7 knots is really in a fresh wind of about 17 knots.

Big, breaking, or irregular waves are tough to sail in. The Beaufort Scale is an especially helpful way to describe weather because it shows both wind and waves on a scale of 13 "forces," from calm to hurricane. This scale reflects the fact that the wind's power increases exponentially. A 22 knot wind is as much as four times more powerful than an 11 knot wind.

We'll take a look at three types of heavy weather: Strong winds, squalls, and storms. For the coastal cruiser the ability to sail comfortably and safely in a good blow is essential. Otherwise you will live in constant fear of strong wind, and not enjoy cruising. Squalls are also a reality of coastal cruising, particularly in the heat of the summer. Though often short lived, their impact can be devastating. A squall drill will be detailed below to help you prepare. While true storm sailing is a rarity for coastal cruisers it behooves us to prepare for that eventuality. We'll look at how to cope with the wind and waves of a true storm, and also suggest how to avoid storms in the first place.

*Inflatable PFDs are comfortable to wear, and provide up to 35 lbs. of buoyancy when inflated –
more than a traditional Type I, and more than twice the buoyancy of a Type III vest.*

2. Heavy Weather Forehandedness

While a big storm rarely arrives unannounced, heavy weather and squalls can sneak up on you if you're not alert — especially on a run, when you and your crew can be distracted by the thrills of sailing fast, the seas are following, and the apparent wind is relatively light.

Be alert to the rise and fall of the barometer, changing cloud formations, radio forecasts, and changes in the wind, waves, and air temperature or humidity.

If a squall or hard blow seems imminent:

• Plot your position using every means at your disposal, and note the sea room and safe harbors in every quadrant.

• Put on foul weather gear and safety harnesses. You can always take them off if the weather calms. See *Chapter V - Safety*, for more on harnesses.

• Organize the crew, assigning duties appropriate to skills.

• Big wind and waves are noisy. Establish hand signals for communicating. For example, fingers up means *hoist or trim slowly*, thumb up means *hoist away*, or *full trim*. Likewise, fingers down for *ease*, and thumb down for release. A fist means *hold* or *stop*.

• Close hatches and other deck openings to keep water from coming below.

• Secure loose gear on deck and below. Rig lee cloths below. Clear excess gear to reduce windage on deck and in the rigging.

• Be prepared to shorten sail quickly.

• Put your best steerer at the helm, particularly for the initial blast of a squall. Steering is demanding in rough weather.

• Make sure flashlights, tools, and spare gear are ready.

• Prepare some simple food, and drinks with straws.

Leaping off waves – and crashing back down – is tough on the crew and equipment. Speed up, slow down, change course, shorten sail, pick a new destination – do something – to stop the pounding.

Twisted sail shapes, with the upper portions almost luffing, spill power and create a forgiving steering groove when sailing in waves.

3. Heavy Weather Sailing

Sailing well in strong winds can allow you to cover many cruising miles. Not only that, but it is exhilarating. To harness the power of the wind and battle the waves, to keep control and sail well. That is the goal.

Depower

The challenge in heavy weather is to reduce power (*depower*) to keep control, while keeping sufficient power to fight the waves which come with heavy winds - and sufficient speed to get to your destination. The slower you go, the longer you'll be out.

As detailed in *Chapter III – Upwind Sail Trim*, depowering techniques to flatten sails, increase twist, and reduce angle of attack are the first steps in dealing with increasing winds. When the methods described there are not sufficient, then stronger measures are called for. We'll address those here.

The waves which accompany strong winds are as big a problem as the wind itself. Pounding upwind against building seas can be more than unpleasant; it can be dangerous as the motion batters the crew and equipment.

Waves also make depowering tricky, as

If depowering the sails is not sufficient, then shorten sail to match conditions. Here a small jib and deeply reefed main allow for comfortable sailing in strong winds.

sailing underpowered in waves can leave you wallowing, and put you at the mercy of the waves. The challenge is to keep enough power to handle the waves, while still maintaining control.

There are several ways to reduce pounding. First, add twist to your trim, to give a wider steering groove. This will allow you to steer around the biggest waves. Next,

Heavy air trim: With the main and genoa trimmed flat, this boat is fully powered in fresh breeze. The traveler could be lowered to further reduce power before shortening sail.

Rolled and Reefed: Reducing sail area often restores control and improves speed. Set your boat up so you can easily shorten sail, and do so early in a blow. You can always unreef and unroll if you are under-powered.

change speeds. Sometimes sailing faster will smooth out the ride, as you power through the waves. Ease sails a bit, and bear off a couple degrees. Another option is to slow down. If the boat is leaping off the waves, then shorten sail and slow down to keep the boat in the water.

You can also improve the boats motion through the waves by moving weight out of the bow and concentrating it amidships. If you know you're going out in big seas, then move the anchor and rode out of the bow locker, and stow them below, perhaps in a couple of big canvas bags.

Another option to consider is picking a new destination. Do you really need to go upwind in these big waves? Let's reach off and go somewhere else!

Slowing Down and Shortening Sail

If depowering alone is not sufficient, then it is time to reduce sail area. In heavy winds a well trimmed reefed boat can provide much better speed, control, and comfort than an over canvassed boat. Start

by reefing the jib. Generally, a partially rolled jib and full size main is more effective than a reefed main and big genoa. If you are still over powered, then reef the main.

As mentioned above, sometimes *slowing down* a little can dramatically improve the motion and comfort of the boat. At other times, adding power and speed sufficient to overmatch the waves can improve the ride. Often adding twist by easing sheets just a couple of inches will help the boat find a better path through waves. If the motion is bad then experiment to improve it.

A Smaller Jib

The first step in reducing sail area is to reduce jib size. With modern materials, sail strength is less an issue than size and shape. The sail will be wrong from a performance standpoint before it will be damaged by the wind. [This statement is not a warranty. Do not blow out your sails.]

Depending on your set up, you can reduce jib size by changing to a smaller sail or roller reefing your genoa.

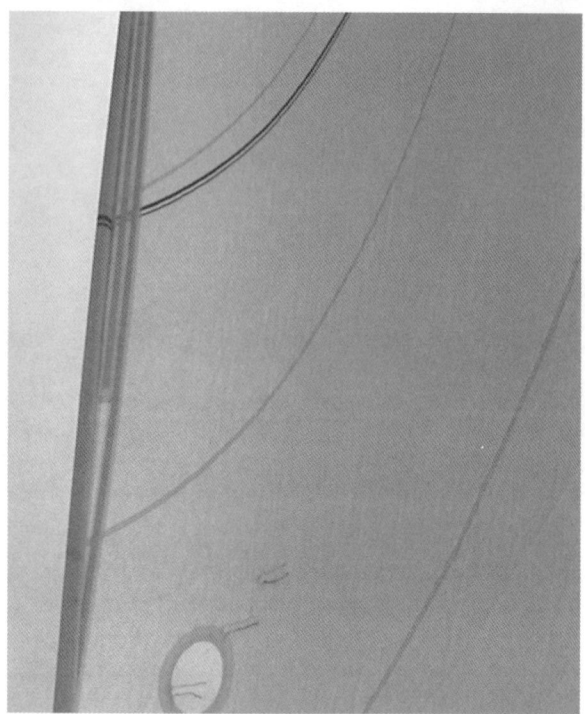

A foam or rope luff helps maintain good sailing shape in a roller reefed sail.

For roller reefing mark the foot of your genoa at increments and put corresponding marks on your jib tracks for proper lead position.

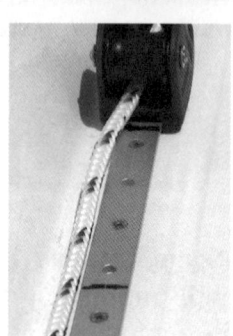

Roller Reefing

While roller reefing can reduce sail area, roller reefed sails are a compromise in sail shape; though foam or rope luffs and other refinements have vastly improved sail performance. To protect the life of your sail be sure to leave a portion of that tack patch exposed to handle the loads along the foot.

As the genoa is rolled the jib leads will need to be adjusted to keep proper jib shape. Marks on the foot of the genoa for first and second increments of rolling – after perhaps 3 and 6 rolls on the headstay – and corresponding marks on the jib track for proper lead position can take the guess work out of resetting the leads for heavy air sailing.

A sail inventory which includes a full size light air genoa and a smaller working jib can provide a great boost in performance, control, and comfort, albeit at the cost of requiring an occasional sail change, and room to stow whichever sail is not rigged, not to mention the cost of cost!

Two Jib Inventory
Small Jib, Big Main

For upwind performance a smaller jib and full size main is preferred to a big genoa and reefed main. The small jib, big main sail plan provides better speed, higher pointing, and more control in waves or gusts.

Change to a smaller jib early – as soon as the thought occurs to you – and while it is still relatively easy to do. Change while at the dock or at anchor if you anticipate a breezy day. A smaller jib presents only a small compromise in performance in moderate winds, and keeps sailing comfortable and fun in heavy air.

If still overpowered, then reef the main.

• *Assume a close hauled or close reaching course.*
• *Ease the mainsheet and vang.*
• *Lower the halyard and secure the reef tack (top photo).*
• *Re-tension the halyard.*
• *Take up hard on the reef line – it should pull down and out – like an outhaul, to pull the reefed sail flat (second and third photos).*
• *Trim the vang and mainsheet.*
• *A mark on the halyard and a mark on the mast can help in setting the reef (bottom right).*
• *Tie the reef line snug to the boom (bottom left).*

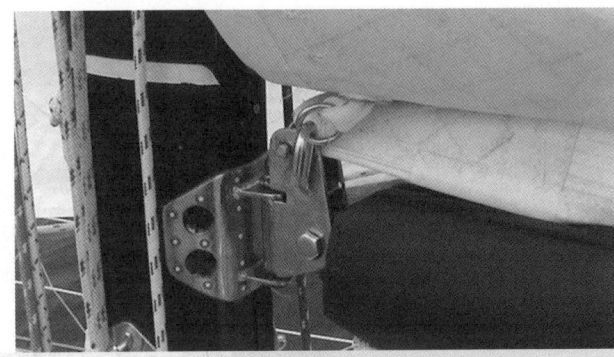

Reef the Main

There are a number of effective reefing arrangements, but there are common elements to each. The first is ease of use. Easy to set and shake. Second, the reefed sail must have a shape appropriate to the conditions, which means flat. The reefing system must pull the sail out, like an outhaul, to keep it flat, as well as pulling down.

When reefed, all the load should be on the reef tack and reef clew. The reef points along the belly of the sail are used to tie up the loose body of the sail, but they should not carry any load. Use colored sail ties and remove them before you shake the reef.

Single or double line systems can both be set up to allow you to reef without leaving the cockpit. Alternately, you may have to go forward to set the reef tack on the tack horn.

Regardless of the particulars, it is important that your system work well, so you are not reluctant to use it. When in doubt, reef. If it proves wrong, shake the reef.

Here are some details on reefing:
• Set the auto pilot to steer closehauled under jib alone, or heave-to.
• Release the mainsheet and vang.
• Lower the main halyard. Mark the line so you know how far to lower it. Pull slack out of the reef line so it won't tangle.
• Secure the reef tack.
• Tighten the halyard.
• Take up very tight on the reef line.
• Snug the vang and trim the mainsheet.

If you expect to be reefed a good while, tie down the reef clew with a sail tie to take the load should the reef line fail.

< proper
not good >

Running wing and wing before a good blow sure beats going the other direction. The boom should be trimmed with a preventer and vang, and the jib should be trimmed through a pole which in turn is set to a topping lift, foreguy, and afterguy.

Downwind in Heavy Air

With the breeze aft you can cover many miles under jib and main, sailing wing and wing, with the jib poled out, and the boom secured by a preventer.

Sailing in strong winds requires strong gear. You need a spinnaker pole (not a whisker pole) supported by a topping lift (or spare halyard) and secured by an after guy and foreguy. The genoa sheet should run through the end of the pole. The pole is not attached directly to the clew of the sail.

The mainsail is rigged with a boom vang to control twist, and a separate preventer, run forward, which holds the boom out. In rough weather the preventer should run from the end of the boom through a block forward on the bow, and then aft, to be controlled from the cockpit. A preventer vang combination, lead to the toerail, can cause problems, particularly as seas build and the boat rolls. If the boom dips into the seas the pressure of the water against the preventer can bend or break the boom, or cause the boat to spin out of control.

One danger of a preventer is the false sense of security it can provide. As you wander on deck do not assume that the preventer will stop the boom from jibing. *Always* keep your head low.

In rough weather rig the preventer from the end of the boom to a block on the bow, and then aft to the cockpit.
>>

Surfing down waves with the wind at your back is one of sailing's greatest rides. Proper trim and careful steering make it exhilarating rather than hair-raising. ∨

Top: With the spinnaker to weather the boat rolls and heels to weather. This can lead to a jibe broach.

Middle: This spinnaker is flying too high. Note how the upper quarter is nearly horizontal. This spinnaker will wander from side to side, and cause the boat to roll from side to side.

Bottom: This spinnaker is choked down, and under control in very strong winds. The spinnaker pole is lowered, and the sheet overtrimmed to put the spinnaker directly over the bow. While you may not want to fly your spinnaker in this much breeze, you can employ the same techniques to keep control in more moderate conditions.

Spinnaker Sailing

Under spinnaker in heavy air – are you crazy? It is in these conditions that the Snuffer techniques described earlier (*Chapter III*) are most valuable.

If not properly trimmed in heavy winds, the spinnaker can overpower the boat. But through better trim you can increase the wind speed under which you feel comfortable with a spinnaker, and raise the threshold at which you can carry the sail and maintain control.

In heavy air, if the spinnaker rolls out to weather, then it can roll the boat too. In more moderate conditions it can work well to square the pole aft, and get the spinnaker out from behind the main. But in heavy air this same trim creates problems.

Likewise, if the spinnaker flies too high and far from the boat, then it will wander from side to side. As the spinnaker swings, the boat rolls, making steering difficult and control tenuous.

To keep control of the boat you need to control the spinnaker. Choke it down directly in front of the boat by lowering the pole and over trimming the sheet. This "short leash" prevents the spinnaker from wandering and pulling the boat out of control.

Sailing under Gennaker in fresh breeze will let you tick off the miles in a hurry. Ease the tack line slightly, and ease the sheet to curl the Gennaker luff, retrim to take out the curl, and then overtrim a foot or two to provide more latitude in steering. You will need to steer actively to keep the boat on course as the waves and sails try to round the boat up or roll it down.

Gennaker Sailing

Sailing under Gennaker in fresh breeze and following seas requires careful attention to trim and steering. Ease the tack line and gennaker sheet to allow the gennaker to roll out from behind the mainsail, and steer actively to keep the bow pointed "downhill." To prevent a broach in gusts ease the gennaker sheet. If things get out of control, snuff and stow the gennaker and switch to a jib.

Tacking and Jibing

The waves which come with big winds can make basic maneuvers challenging. When tacking, look for a relatively smooth spot, and start your turn as the bow climbs a wave. Push the helm over so that the next wave will push the bow down onto the new tack.

In extreme seas you may not be able to tack at all, and may instead need to *wear ship* or jibe.

Of course, jibing in heavy air is no picnic. Often the best way to handle the jib is to roll it up. A heavy air jibe is best accomplished at speed. As the boats surfs down a wave loads on the sails are reduced. Use extra hands to jibe the main, and ease it quickly. Watch your course and steer to control the boat as it tries to round up coming out of the jibe.

Once under control you can unroll the jib. Put the roller furling line on a winch for control while easing, as the load will be too great to handle bare-handed.

Motor Sailing

Perish the thought! This is a sailboat! *Well yes, but we're not racing!*

If you are sailing under reefed main and rolled genoa and you are still over powered, stow the jib and crank up the "iron genny."

Motor sailing into wind and waves with the main set provides a much better ride than motoring with no sails. (Save that for days with no wind.)

Motor sailing lets you point high, and make great progress to windward without the violent pitching of motoring into seas with no sails set.

Trim the main, head up high enough to control your angle of heel, set the autopilot, and keep a lookout.

Make sure cooling water is pumping through the engine. On some boats the water intake lifts out of the water when heeled. Violent pitching can also allow air into fuel line, which stalls the engine, and requires a bleed. A further difficulty is that the pitching boat may stir sediments off the bottom of the fuel tank, which can in turn clog the fuel lines or fuel filter.

Motoring with *no* sails will probably *not* work - particularly in big seas. Sails are needed - at least a reefed main - to provide some stability and extra power.

Also to be avoided is motoring across a beam sea, as that can lead to violent rolling, or even a broach.

4. Squall!

A summer storm can hit with stunning suddenness, turning a languid late afternoon into a trial. While often short lived, a squall's sudden arrival requires a quick response.

Squall Drill

A little preparation goes a long way. Here are some things you should do when you first expect a squall, and when its arrival in imminent.

- Don lifejackets and harnesses (if you don't wear them habitually).
- Have foul weather gear at hand.
- Clear loose gear from the deck.
- Close open ports, prepare the hatch boards, and secure loose gear below.
- Plot your position by every means available.
- Determine where hazards and safe water lie.
- If time allows, head in, but beware:

The *worst* place to be when a squall hits is *almost* in, caught in a constrained space amidst a crowd of boats all dashing home. Certainly the preferred place to be is secured in your berth or mooring. The next best place is in open water, clear of crowds.

- Prepare to shorten sail. One approach is to take a deep reef in the main at the first hint of trouble, and then to roll the jib completely with the first cool gust. Make sure the jib reefing line is ready to go, with clean wraps on the furling drum.

When the squall hits, you want to be in full foul weather gear, harnessed to the boat, with the boat buttoned up. You want to be under reduced sail, with plenty of sea room, and you want to be sailing to safe open water.

Two other suggestions: Practice your Squall Drill in benign weather to see how quickly you can shorten sail, and pay attention on sultry summer days, so you won't be caught off guard.

To recover from a broach, hold on, and ease the sheets. You may need to ease the vang as well to get the boom out of the water.

It is not just storm wind, but storm waves which make heavy weather sailing so challenging.

Broach

If you are caught with too much sail in a sudden squall then the boat may broach before you can shorten sail. In a broach the boat is laid over on its side by the wind. It can take eternal minutes to bring the boat up, and there are dangers to address in your response.

First, hold on, and take your time. Though there is much sound and fury, there is not as much danger as it might seem, as long as every one holds on, and stays aboard.

In order to stand the boat up you will need to ease sheets. The jib sheet may be difficult to get to, as the winch may be awash to leeward. Likewise, the mainsheet will be heavily loaded and difficult to release. Even with the mainsheet eased the sail may not spill, as the boom may hit the water, preventing it from running out. The boom vang will need to be eased as well.

The greatest danger on deck may be from falling across the dramatically heeling decks, while below there is a danger of being thrown across the cabin and of being pelted with anything which is not secured.

As the boat comes upright beware the flogging sheets, which can whip with remarkable force. Shorten sail and survey the boat above and below for gear which may have fallen or shaken loose.

5. Storms

As compared to the quick response and sudden nature of a squall, sailing through a storm in open water is an endurance contest. In addition to wind, we need to deal with waves, and fatigue.

Sailing in Waves

Sailing in big waves is a test of seamanship and steering, which is why you should put the best steerers on the helm. Experienced dinghy sailors often are very good at it because they see "survival" weather more often than most cruisers.

Avoid sailing on a reach across tall breaking waves. They can roll a boat over.

When sailing closehauled in waves, aim toward flat spots while keeping speed up so you can steer. Tack in relatively smooth water. Waves washing across the deck carry tremendous force. A cubic foot of water weighs 64lbs. A wave can bring many hundreds of pounds of water across the deck.

Sailing on a run or broad reach in big waves is exhilarating, but be careful not to broach and bring the boat beam-to a breaker. Rig a preventer to hold the boom out.

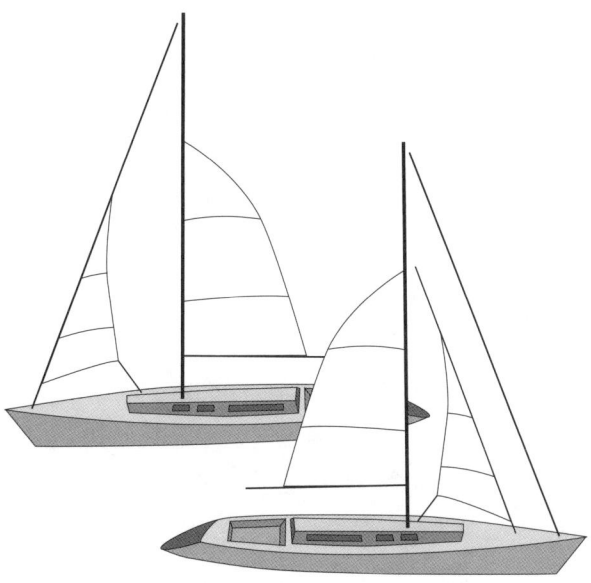

The storm jib can be rigged on the headstay or on its own inner forestay.

Storm jibs (above) and storm trysails (bottom) look laughingly small until you put them up in a storm.

Don't wait until you are caught in a storm to figure out how to rig and lead the sheets for your storm jib.

The storm trysail rigs on its own track adjacent to the mainsail groove.

Storm Sails

You can also shorten sail by setting storm sails — the storm trysail and storm jib. They may seem tiny but because wind force rises exponentially, they're the right size. Storm trysails are usually trimmed to the rail, but some modern ones are set on the boom. The storm jib should be set near the mast to keep the sail plan's center of effort near the boat's center of lateral resistance. This helps the boat balance.

Sailing Strategy

The first decision before an approaching storm is the toughest: Run for cover, or head out for sea room. With modern forecasting, a true storm will rarely arrive unannounced. But as you venture further offshore the chances of being caught out increase. While running for cover seems the preferred choice, the danger lies in being caught in the storm, close to shore, with no room to maneuver or run off.

Two classic storm *strategies* are to try to keep away from land so you're not blown up on shore, and to sail out of the storm's path — especially its "dangerous semicircle," which is its right side as it advances.

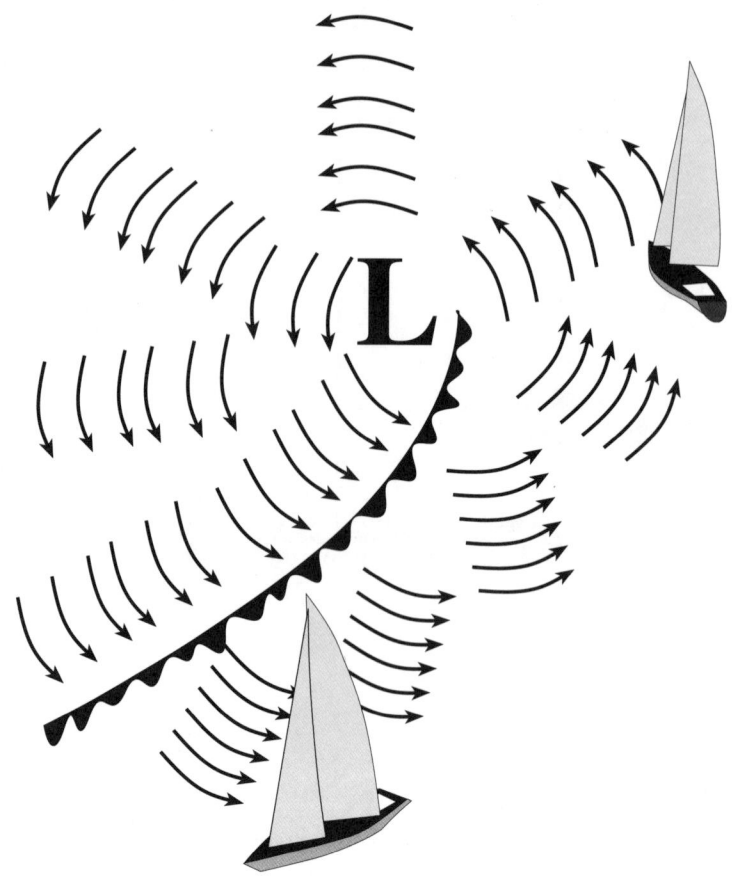

Storm Strategy: The best place to be is in a protected harbor. The worst place to be is almost in a protected harbor – running for home and getting caught near shore, with little sea room. If caught out to sea in the face of an oncoming storm, try to escape the dangerous semicircle on the advancing right side.

Storm Tactics

Storm *tactics* are ways to handle a storm once you're in it. There are several proven choices, all of which aim one of the boat's ends (the bow or stern) toward the waves. No one tactic will work best for all boats in all conditions.

- Sailing under storm jib and deeply reefed mainsail or storm trysail provides the most control. Sails give you the power to steer and control your boat in the waves.
- Running before the storm with the stern toward the waves, perhaps towing a drogue to slow the boat. The boat must be steered actively. This tactics depends on sea room. Another concern is remaining in front of an approaching storm, rather than sailing out of its path.
- Heaving-to on a close reach with the jib trimmed to windward. Heaving-to can be

an excellent heavy weather tactic, though some boats fare better than others. More details follow on the next page.

- Lying to a sea anchor while hove-to or under bare poles. A sea anchor is a small parachute set at the end of a line off the bow. A sea anchor helps keep the bow up into the waves so the boat won't end up beam to the seas. One concern is the load on the rudder as waves push the boat aft.
- Another alternative is lying ahull, simply sitting with sails down. This passive alternative is less reliable than the other tactics, as you lose the ability to control your angle to the waves, and may end up beam to the seas. Furthermore, the motion of the boat rolling in the waves without benefit of sails can be further debilitating.

If caught in a storm several techniques can help you weather it. Modern designs generally fare better with more active tactics.
Hove-To, with the jib backed, main trimmed hard, and helm down, the bow will stay up while the boat crabs to leeward.

Heave To

Wouldn't it be great if, during a heavy air sail, you could just take a break, and relax for a bit? Imagine a short respite from the relentless pitching and pounding. A chance to rest, take a meal, or check over the boat in relative tranquility. Well, you can.

The lost art of heaving-to allows you to "park" in open water. Hove-To trim has the jib trimmed aback (i.e., to the wrong side), the main trimmed hard, and the helm lashed down. Trimmed this way, the jib tries to push the bow down. As the bow turns off the wind the main fills, and the boat moves forward. With the helm lashed down, the rudder turns the boat toward the wind. As the main goes soft the jib once again takes over, pushing the bow down. The main refills, and the rudder pushes the bow into the wind again.

Achieving this balance will require some fine tuning, depending on the wind strength, your boat design, and the sails flying. You may, for example, need to roll the genoa most of the way, to match the wind strength. Also, modern fin keeled boats do not heave-to as well as more traditional designs.

The boat won't actually stop. It will lie about 60 degrees off the wind, sailing at 1 or 2 knots, and making leeway (sliding to leeward). The motion will be much less than under sail, and dramatically more stable and pleasant than dropping all sails and lying ahull.

In storm seas some boats benefit from a sea anchor set off the bow to help hold the boat up into the waves.

Modern boats often heave-to best under a small jib trmmed well to weather. Finding the best set up for each boat and set of sailing conditions requires testing.

A partially rolled genoa and full main balances well for this boat in moderate conditions.

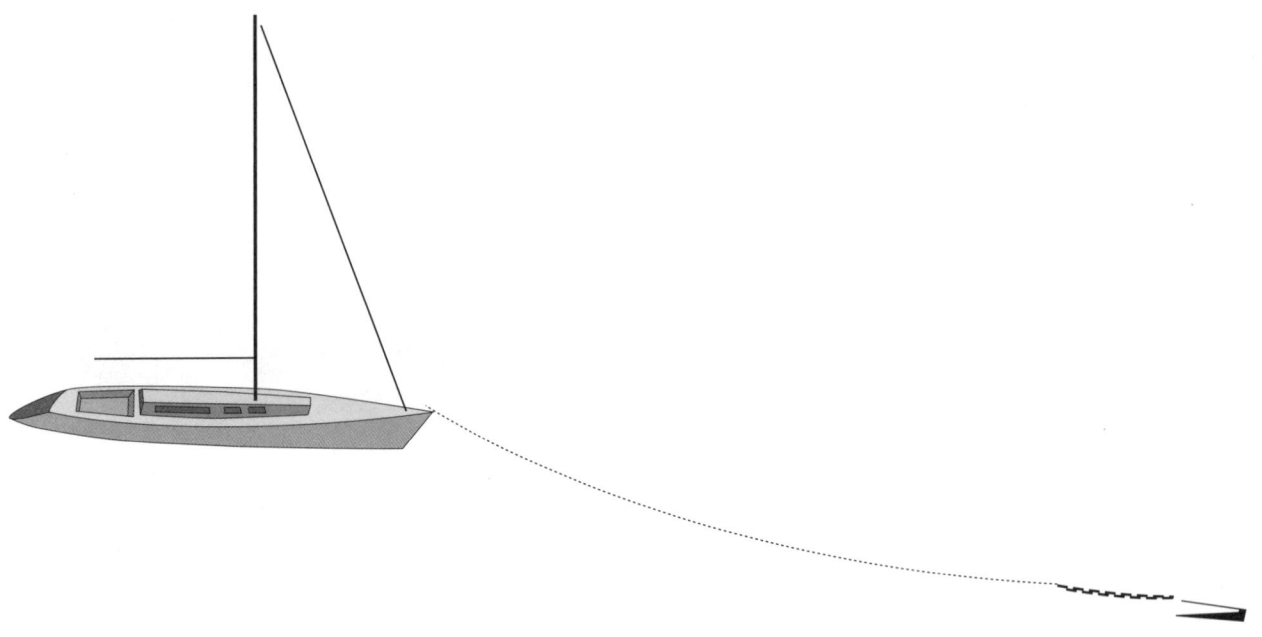

6. Alternate Storm Strategy

Another option in storm strategy is… don't go. If you are caught in a storm you'll have to endure it, but no one came on the cruise to do that. If conditions are wrong, or are forecast to worsen, don't go. If you can avoid the storm, then do so.

If your at home, stay in the slip. If you're mid cruise, button up the boat, and read a book or play cards. Relax. Enjoy the time with your shipmates. Study the pile of *Owners Manuals* you've accumulated with each piece of new gear. Tinker with boat projects.

Put some soup on the stove, and peak out every so often to see that you aren't dragging. Shake your head as you return below, and remark, "My oh my, is it nasty out there."

If your boat is threatened by a tropical storm or hurricane, then strip all excess gear from the deck, double up all docking or mooring lines, protect those lines from chafe, and get off. Don't risk you life to save your boat.

7. Misery and Danger

Although everyone will remember it differently years later, a long, wet, cold sail through a storm can be miserable.

As a skipper you can make the best of it, by watching over your crew, and by offering relief or help to those who need it, and encouragement to all.

"This is miserable, but it will end."

Take the time to marvel at the forces of nature, and at your ability to carry on in the midst of the storm. Few people get to experience the full fury of a storm. It may not be pleasant, but it is memorable.

While misery and discomfort can lead to fatigue, diminished performance, and danger, do not mistake one for the other. Distinguish in your own mind the difference between misery and danger. Don't attempt a dangerous harbor entrance to escape misery. Don't compromise the safety of the boat and crew to escape discomfort.

A further discussion of related safety issues is upcoming, in Chapter V, next.

Chapter V – Safety

1. Risk
2. Formula for Disaster
3. Prepare the Boat
4. Prepare the Crew
5. Prevent Emergencies
6. Emergencies!
7. Equipment List

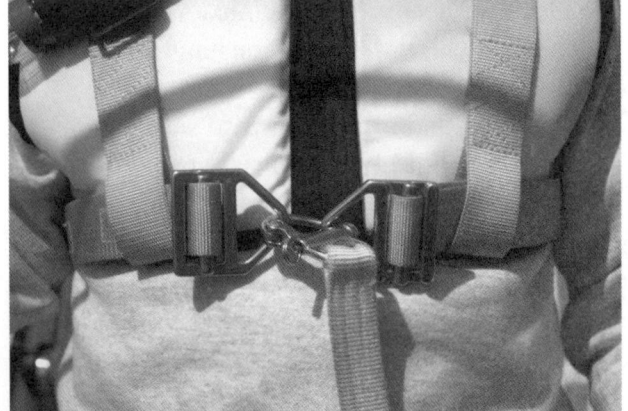

Safety

1. Risk

There are risks in going to sea – and risks in remaining ashore. The first step in dealing with danger is to accept, rather than deny, it. While inescapable, the way we prepare for and handle risks is a central factor in how and whether we survive.

Once we accept risk we must assess and prepare to deal with the various dangers we face. That process includes procuring, maintaining, and practicing with safety gear.

Another central facet of risk management is prevention. As the old Navy expression states so well, "The price of safety is eternal vigilance." It is our own efforts that prevent dangers from becoming disasters. By keeping concerns over safety foremost in your mind most emergencies can be averted. While this may not match the romantic vision of cruising, it is a stark reality, and in fact the following of shipboard routines in the name of safety can provide a satisfying rhythm to passagemaking.

When prevention fails we will need to reckon with emergencies as they happen. Our preparations and practice will put us in the best position to deal with emergencies successfully.

Stop, Look, Listen

As a rule - slow down. Assess the situation, and consider your options. Rarely is a hurried response required, or best.

True Emergencies

There are a few true emergencies which require immediate and appropriate response - crew overboard, sinking, and fire. For these we will need to be particularly well prepared.

2. Formula for Disaster

As detailed in *The Annapolis Book of Seamanship**, there are seven factors found in most catastrophes. It is not any one thing that creates disaster, but the accumulation of several factors, piled one upon the next. Here are the seven factors:

1. A rushed or ill-considered departure

Too often, our decisions on when and where to go are dictated not by the judgment of prudent seamanship, but by factors of time and schedule which do not answer to the sea. The lives we lead, and the schedules we keep, can lead to compromises in safety. Rushing to get home ahead of an approaching storm, rather than sitting tight in a snug harbor, is an easy error when sitting tight means missing a scheduled rendezvous or a timely return to work.

2. The route is dangerous

Areas close to shore can be as dangerous (or more so) as routes in open water. While open water sailing may leave you exposed to wind and waves, the dangers near shore can include narrow channels, strong or unpredictable currents, traffic, and a lack of sea room. Time your travels to passage dangerous areas at the least dangerous times. For example, in areas of strong tidal currents, time your passage to transit the worst areas at slack high or low.

---

* *The Annapolis Book of Seamanship*, page 342.

3. The route has no alternative where the crew can "bail out"

As you plan passages consider intermediate destinations where you can seek refuge. Study your options in advance. As you pass each, consider whether it might be prudent to pull in. If conditions deteriorate, keep in mind that it is sometimes best to turn back, rather than press on.

4. The crew is unprepared

The skills and capabilities of the crew must be up to the challenge of the passage. A shorthanded or inexperienced crew can be pressed to its limits by the challenges of a routine passage, with no reserves for the unexpected. Our passion for adventure and search for new challenges must be tempered by the possibility of the need to rise to a challenge beyond the scope of those we anticipated. Preparation of the crew by training for emergencies - such as crew overboard rescue or squalls - can help overcome limited experience. But in the end one must make sure the crew is sufficient in both number and experience. The crew must also carry appropriate personal gear to protect themselves from the elements.

5. The boat is unprepared

The boat and its gear must match the passage. Of course the design of the boat itself must match the route. Some boats simply are not intended for open water sailing. Beyond that is the matter of equipment - from blocks and lines to safety and navigation equipment. You've got to have the right stuff, take care of it, and know how to use it.

On the following pages you will find lists of gear matched to the degree of self sufficiency required for coastal cruising.

6. The crew panics after an injury

An injury to one crew member can lead to a cascade of bad decisions which threaten the well being of the entire crew. Psychologically, the feeling that everything will be OK may be shattered, and fear takes hold. Often there are also misguided decisions based on a false sense of urgency, when the injured crewmate's condition could be stabilized without a rush to shore, and danger.

7. Leadership is poor

In times of challenge the crew will look to the skipper for leadership - strong, well reasoned decisions which draw from the skills and knowledge of the entire crew. A brash, irrational, or defensive skipper can quickly lose the confidence of a crew, and the chain of command will then collapse.

Prepare and post a schematic of your boat showing where safety and emergency gear is stowed, and showing the location of all through hull fittings and tank valves.

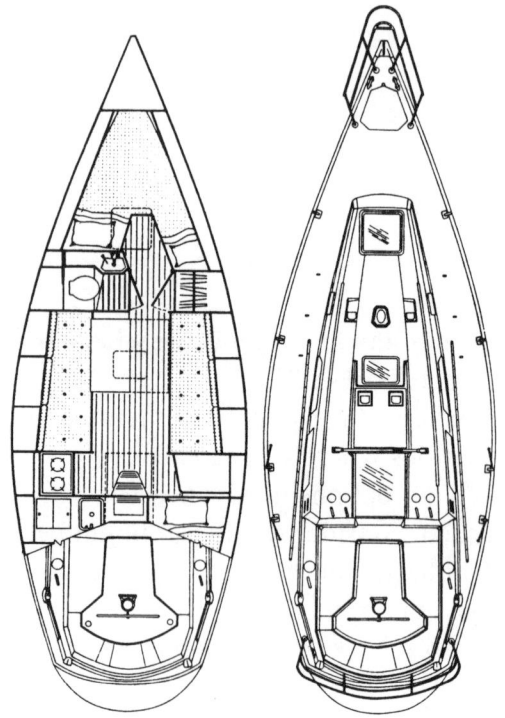

3. Prepare the Boat

Your safety depends on your boat. The care and use of your boat's sailing gear can help assure that you'll not need to use your emergency gear. Obviously it is not enough to just buy the gear. You need to install it, maintain it, and know how to use it.

In addition to your sailing gear, you must have safety gear appropriate to your sailing. You cannot protect yourself from every eventuality. Select gear to match your sailing, and when choosing between items, select the simpler, easier to understand, and easier to use device which solves most of the potential problems rather than the more comprehensive, complex, and harder to use alternative. As you make your selections try to imagine what it would be like to use the item in the most trying of circumstances.

You also must keep you boat in working order. Your sails and sailhandling gear must be up to the task you will put before them. Worn or damaged gear is an invitation to failure, and with it, distractions and further troubles.

Psychologically, if the crew lose confidence in the boat, fear can cloud the judgment of the crew in dealing with problems as they arise.

In this section we'll consider gear in four categories: US Coast Guard Regulations, Installed Boat Safety Gear, Personal Safety Gear, and Emergency Gear. At the end of the chapter are more detailed lists of specific items and ideas.

US Coast Guard Regulations

The Coast Guard is very specific as to what's required in the way of flares and fire extinguishers. Coast Guard Regulations also cover horns, running lights, and lifejackets, among other items. Review that list annually and make sure your gear is up to date. Safety authorities strongly recommend that the flares be manufactured to SOLAS standards – those are the standards established by the Safety Of Life At Sea convention, the international body that governs all maritime safety. SOLAS flares are more expensive, but they're also much more effective. They burn brighter and the parachute flares have a much longer hang time than the conventional alternative.

Installed Boat Safety Gear

This means engines and their gear, bilge pumps, and through-hull fittings. All must be tested regularly and checked for problems. Heavy gear should be secured in place. You should have one big manual bilge pump as well as an automatic electrical pump. Through-hulls fittings should be inspected regularly and have soft pine plugs nearby to plug the holes if they or their hoses fail.

Personal Safety Gear

This means lifejackets or PFDs (the number of which is mandated by the Coast Guard) as well as flashlights, whistles, knives, and safety harnesses – plus foul weather gear (because a dry sailor is a warm sailor, and a warm sailor is a safe one).

We use and strongly recommend inflatable lifejackets. When they're not inflated, they're far more comfortable and far less awkward than the usual kind with fixed buoyancy. When they're inflated, they have far more buoyancy.

The normal body needs about 8 pounds of buoyancy to keep the head afloat. A Type III PFD (Personal Flotation Device) has 15.5 pounds, which is not enough to keep most people's faces out of the water if they're unconscious or exhausted. A Type I has 22 pounds of buoyancy, which will keep your head clear – yet the Type I is extremely bulky and awkward to wear. Now, many inflatable lifejackets (which are far more comfortable to wear than a Type III) have even more buoyancy than a Type I – up to 35 pounds of buoyancy – enough to keep your head clear of waves.

When should you put your lifejacket on? That is the wrong question. Put it on, and ask: *When should I take my lifejacket off?*

We also wear safety harnesses a lot and recommend that you do, too, when sailing at night, when shorthanded, or in rough weather. A harness has a tether – a 6-foot

An inflatable lifejacket is comfortable to wear, and provides tremendous buoyancy when inflated. When should you put it on? That is the wrong question. Put it on, and ask, "When should I take it off?"

length of webbing – with a snap shackle that is hooked to the boat or to a long length of webbing or line, called a jack line, that runs fore and aft and allows you to move around. Some excellent safety harnesses are part of inflatable lifejackets – a terrific combination.

You'll find more information on jack lines a few pages ahead in this chapter.

Emergency Gear and Offshore Gear

Some things concern average sailors and ocean sailors alike – like recovering someone who's fallen overboard. We like the LifeSling rescue device, whose operations we describe in a moment.

But if you head offshore across a large lake or part of an ocean for three days or longer, you should carry a life raft and prepare a bag of gear, water, and other items to take with you if you have to abandon ship. Life rafts are expensive and bulky – and necessary if your boat goes down and you have no interest in going down with it. But remember not to expect miracles of them. They're not boats. We won't talk too much more about them here.

4. Prepare the Crew

Your safety depends on your crew. A well prepared boat in the hands of a poorly prepared crew is not safe.

Your crew need to know how to handle the boat and sails – how to rig, and reef, and how to respond to an imminent squall.

They need to know how to operate the ships systems, both for their own safety, and to keep from damaging the boat and her gear.

The crew need to know the location and operation of safety gear – from fire extinguishers and radios to crew overboard rescue equipment and medical stores.

They need to know how to take care of themselves – how to dress for all kinds of weather, how to move around the boat in rough conditions, and they need to know to ask for help with things which are unfamiliar.

As a skipper you need to prepare your crew in a manner appropriate for the intended sail. Of course, this preparation varies by degree, from basic precautions for a short daysail with a group including non-sailing friends, to more thorough preparations of the team planning to venture off-shore.

For a short daysail with non sailors, there are a number of things you can do to make the experience safer and more rewarding for your guests. These ideas are covered in *Chapter VIII - Sailing to a Destination.*

In preparing for an extended cruise, more rigorous preparations are called for. Here are some things you should do or consider:
• Prepare for emergencies through drills. We'll cover this in more detail in its own section, below.
• Discuss the chain of command in the event that you, as skipper, are injured.

• Practice sailing at night. You'll need to get accustomed to sailing with all your senses as your eyes provide less information. You can add tactile information - such as a whipping on a halyard for a reef drop, where a visual marker would serve by day. Night sailing can provide a margin of safety, as you make the most of fair weather. Night sailing is also one of the great pleasures of cruising. The extra caution required is often rewarded by extraordinary beauty.
For a more thorough discussion of the challenges and benefits of sailing at night see Chapter 8, Section 4.

5. Prevent Emergencies

Prevention is of course our goal. A well found yacht, and a properly prepared crew, are key elements to prevent emergencies before they happen.

Prevention takes many forms. As discussed, matching the boat and crew to the challenges of the cruise are an essential foundation. From there, the next step is to establish shipboard routines.

Among the routines are:
Cleating, coiling, stowing and handling line in a consistent way.
Regular housekeeping to keep lines organized and gear stowed.
Habitually checking overboard to make sure no lines are dragging before starting the engine and engaging the prop.
Navigating redundantly through dead reckoning, piloting, and electronic means; and logging navigation data routinely.
Regularly checking weather conditions and forecasts.

When going aloft use a bosun's seat which will hold you securely. Use two halyards, and do not rely on shackles.

The seemingly mundane - like proper and consistent coiling, and regular inspection of running and standing gear, are all part of the "eternal vigilance" required to assure a safe passage.

Periodically surveying the boat for signs of chafe, wear, or damage – at the turn of each watch, for example. Rings and pins holding blocks and lifelines should be inspected, and running rigging checked for chafe.

Monitoring the condition of batteries and charging as necessary.

Routine wearing of lifejackets and use of harnesses.

Periodic drills in preparation for squalls, crew overboard rescue, fire, and sinking.

Practice sailing in close quarters on the expectation that the engine may fail.

Routine maintenance of the engine so it won't fail.

Venturing aloft periodically to inspect the rigging and spreader tip pads, to make sure antenna mounts are secure and to lubricate sheaves.

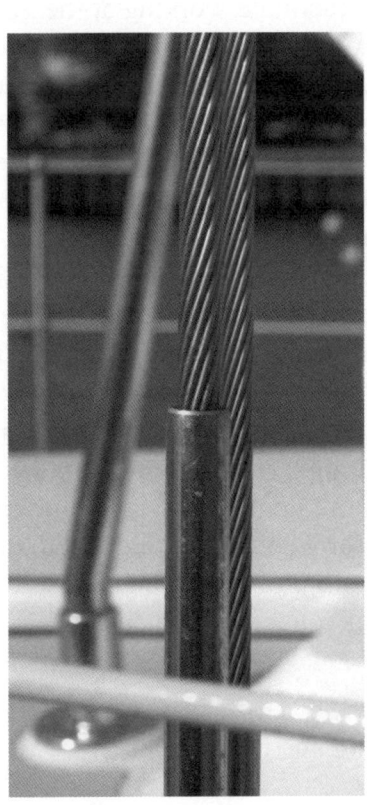

Harnesses and Jack Lines

Stay on board. In rough weather, at night, and when short handed, you should be attached to the boat. This is accomplished through the use of safety harnesses. At the start of your cruise issue harnesses and lifejackets fitted to each crew member. This gear does little good if left buried deep in a locker.

But it is not enough to wear a harness. There must be places to hook them. The lifelines are NOT such a place. Strong, backed pad eyes should be mounted in the cockpit near the helm and trimming stations.

In addition, jack lines of strong tubular webbing should be run the length of the boat on either side of the deck. This will allow a crew member to move the length of the boat without unclipping. Lead them inside the shrouds to facilitate working at the mast, and to allow going forward on the leeward side, if necessary. Often jack lines can be secured to the bow cleats. The aft end should be anchored far enough forward that a crew dragging overboard will hang to the side, and not drag astern.

Encourage new crew to don their harness in moderate weather and accustom themselves to moving about while secured to pad eyes or jack lines. A little familiarity can make a huge difference when venturing forward at night in heavy air.

One more item: While jack lines made from tubular webbing are preferred, they can degrade from prolonged sun exposure. Stow them when you leave the boat.

Jacklines, of webbing, secured at the stem, (above), and of wire, (below) shown here from the bow looking aft, allow a crew member to hook on and move the length of the boat. Webbing is preferred, as wire tends to roll under foot, and has no give.

A combination safety harness and inflatable life vest overcomes the awkwardness of trying to fit both at once.

Take a break. If things are happening too fast – if you can't keep up with navigation, don't have time for a meal, and aren't sure where you're headed, then heave-to. Slow things down, and sort it all out at your own pace before things get out of control. For details on heaving-to, see Chapter IV – Heavy Weather Sailing.

6. Emergencies

While prevention is the goal, training and practice is still a necessity. There are a few true emergencies to address, along with a slew of other problems which deserve attention in their own right, and because they can lead to emergencies if handled badly.

True emergencies require an immediate and appropriate response. They include crew overboard, fire, sinking, and serious illness or injury.

Other problems, while serious, are generally not immediately life threatening, and are often not made dramatically worse by the brief passage of time. Nonetheless, it is prudent to prepare for problems such as going aground, squalls, minor injuries or illness, broken gear and rigging, engine troubles, electrical problems, and navigation difficulties. In addition, there are challenges brought on by weather, sea sickness, food poisoning due to poor refrigeration, hypothermia, heat stroke – the list goes on and on.

Your response in an emergency can make the difference between a bad scare and a disaster. You need to know how to use the safety gear you have, and you need to train your crew in its use as well. Use drills to prepare your crew for emergencies.

Sinking

If the boat is filling fast you have a serious problem. Time is of the essence. Call for help while you still have power.

To prepare for this emergency, test pumps regularly and keep the bilge clean so pumps won't clog. Also, secure soft wood plugs on lanyards near each through-hull fitting. If a through-hull or hose fails, the plug can stop the leak.

Find the leak! A failed through-hull may not be discovered until the floorboards float. If you act quickly, before the water is too deep, the damage will be minimized.

If the hull is breached – say from running into flotsam – you'll know about it. Heel the boat away from the hole to reduce flow. Stuff the hole from inside to stem the flow, then cover the hole from the outside with a sail or other material. Materials placed outside are most effective, as they are pressed into place.

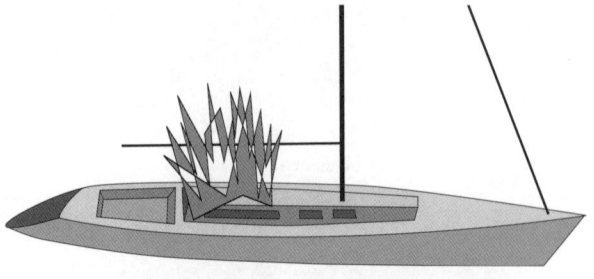

Fire

By law one must carry the number and type of fire extinguishers required by the Coast Guard. Learn how to use them, inspect them frequently, place them in accessible locations where you won't have to reach through a fire to get to them.

Fires occur most frequently in the engine compartment, where heat, oil, and sparks are present. Clean spilled oil out of the engine pan, and check wires for fraying which could promote sparks.

An access port which allows discharge of an extinguisher into the engine compartment, or an extinguisher mounted in the engine box with a remote trigger, can save you from having to open up the box.

Galley fires are the other common type. Cooking oil or grease can spill and ignite. Use baking soda, or an extinguisher – not water – on a grease fire. Also, when cooking under way, wear foul weather pants as protection against burns from spills.

Be cautious near fuel docks. Ventilate well, and beware of spilled gasoline even if you are fueling with diesel.

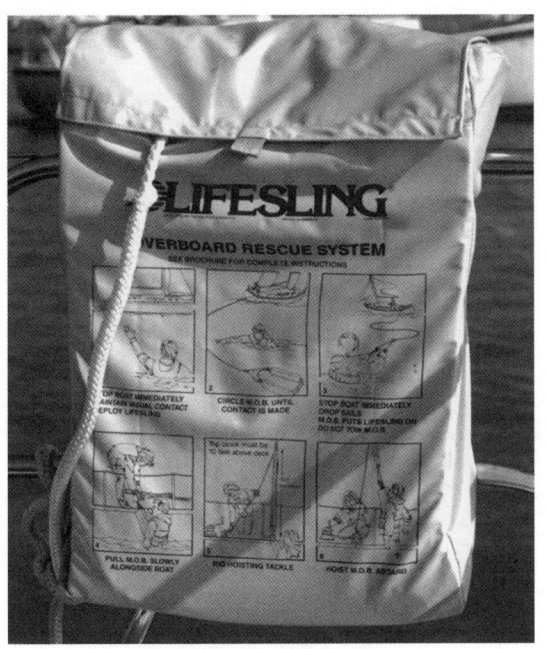

You should own and know how to use – and train your crew in the use of – the LifeSling.
If this is the first time you've seen what's inside the bag, then make a point of deploying it on your next sail, and continue to do so a few times each season so you and your crew are familiar with it.

Crew Overboard

Nothing can shatter the composure of a crew like the cry of, "Man Overboard." It is here, in a situation which screams out for panic, that your preparation and practice will provide the confidence and with it, the composure, necessary to complete a rescue.

We strongly recommend the LifeSling. It is a combination tow rope and lifting sling which has been used with success many times. Successful rescue of a crew overboard (COB) has five steps:
1. Quickly turn back.
2. Get buoyancy to the COB.
3. Make physical contact with the COB.
4. Stop the boat.
5. Get the COB back on board.

If you are under sail when the crew goes over, stay under sail. If you are under power, stay under power. Don't switch modes. Under sail, a quick stop turn can help keep the crew in sight.

The circling maneuver used with the LifeSling has been proven effective over and over for a conscious victim. Practice crew overboard rescue often, and make sure your crew practice too.

1. Immediately tack and stop the boat. Keep the COB in sight and nearby. Don't jibe. A jibe can be dangerous, it eats up lots of distance, and it may leave you with a beat back to the COB. Assign a crew member positioned aft to point at the COB.
2. Deploy the LifeSling. Get buoyancy to the COB immediately. Throw cockpit cushions and life rings – anything and everything to mark the area.
3. Circle the COB. Keep sails trimmed in tight, and circle until the COB gets hold of the LifeSling or sling line.
4. Stop the boat. Under sail or power, come to a halt within a boat length of the COB. (This takes practice!)
5. Get the COB back on board. This can be the most difficult step. Secure the Life-Sling line to the boat. Then use a halyard, or a block and tackle attached to the halyard or the boom, to lift the COB.

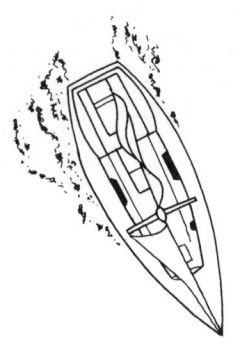

The QuickStop achieves one key goal in our COB rescue: Staying close to keep the crew member in sight. Stop the boat by Heaving-To. From there, deploy the LifeSling and circle the crew member to establish contact.

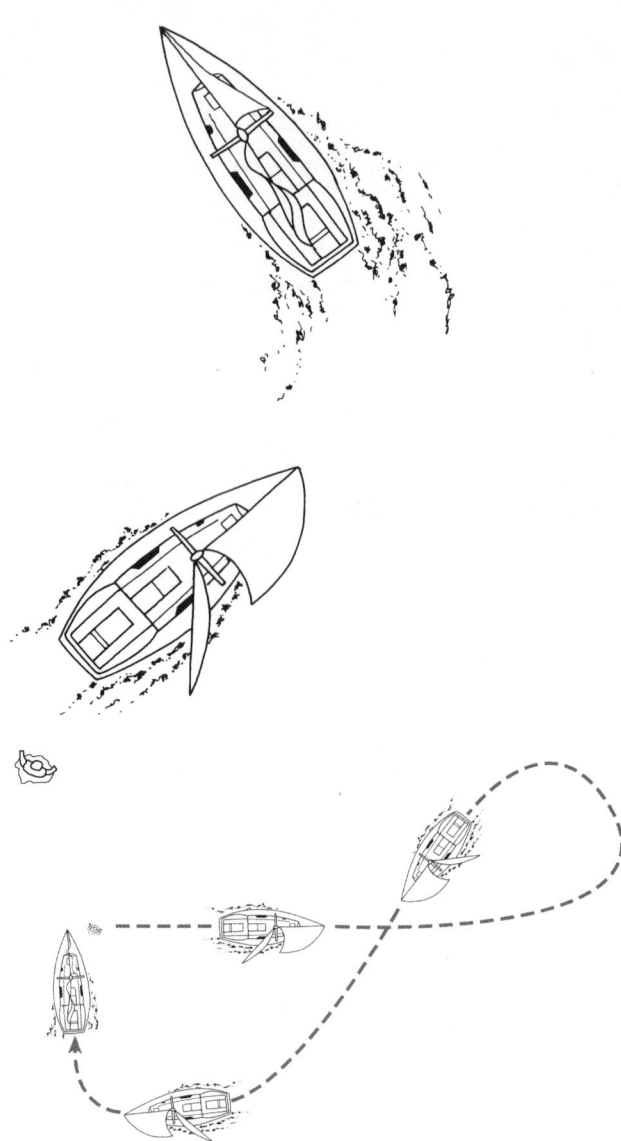

LifeSling COB rescue is simple enough to be described on the LifeSling's pouch, *but you really must practice.*

For most boats, the quick-stop method, using tight circles around the COB, with the sails trimmed tight, is the best way to get the sling to the COB. The reach-and-reach (figure 8) may work better with some boats, though there is a fundamental flaw in that reaching takes you too far away too fast, and you may lose sight of the victim.

Unconscious COB

What if the COB is unconscious or otherwise helpless? You'll have to put another person into the water, and in peril. The rescuer must wear a lifejacket, be attached to the boat with a line, and carry flotation and a second line or a LifeSling to the COB. Each of these lines must be handled by a crew member on deck. Once the rescuer swims to the COB, the lifejacket and second line are attached and the COB is pulled to the boat. Obviously this is very tough with just two in crew. When sailing shorthanded wear an inflatable lifejacket.

The Figure Eight, or Reach-and-Reach is an alternate path for crew overboard rescue. The flaw is that reaching takes you too far away too fast.

You Overboard

If you fall in, make sure somebody knows it! Scream and wave and blow your whistle. Don't waste your strength chasing the boat. Make sure you are seen, inflate your lifejacket, and keep your back to breaking seas. *If you ever fall overboard, you'll feel safer knowing your crew has practiced crew-overboard recovery.*

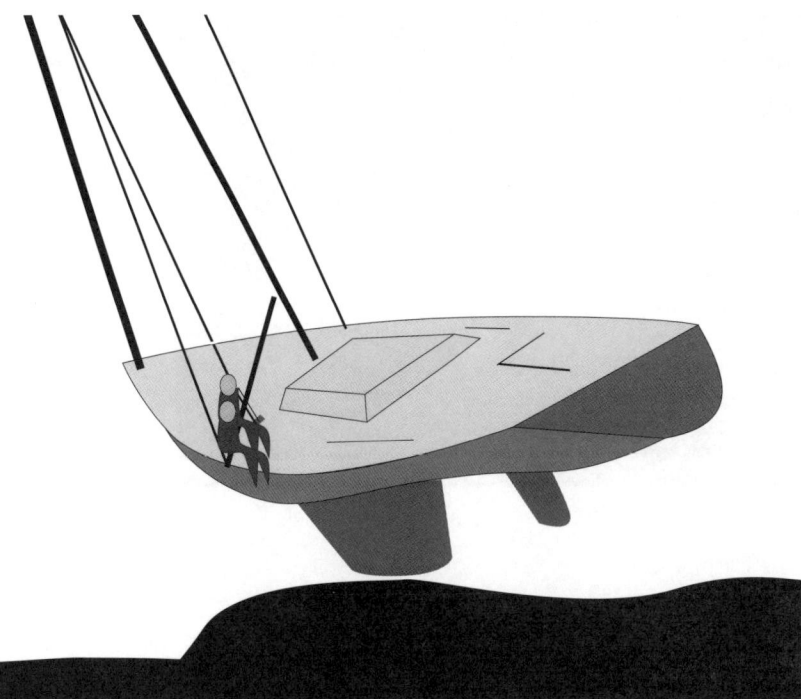

Aground! Turn the boat around immediately, and heel over. You can heel the boat with crew on the boom to lift the keel off the bottom. Be sure to support the boom with the main halyard - not just the boom lift. Failing that, you may have to dinghy out an anchor and kedge off.

Grounding

Quick! Turn around.

While rarely an emergency, running aground can put a crimp in your schedule, if nothing else.

The initial impact of running aground can throw the crew, and cause injuries. When underway in *thin* water encourage crew to sit down so they won't be thrown into a fall on grounding. With good light, a lookout forward can help, though judging depth can be very deceiving.

If you do run aground hard on rocks check for leaks around the keel and keel bolts. Also check to see that your engine hasn't been thrown from its mounts. If the engine runs roughly, the shaft alignment may have been upset.

When you hit, try to spin the boat around immediately, and at the least take pressure off the sails to keep from being driven further aground. You can turn the boat with the sails by pushing out the boom, or backing the jib.

Once you pivot the boat so you are headed toward good water then trim all sails and heel the boat with crew weight to reduce draft. If under reduced sails, you can add canvas to increase heeling - but once again, be sure you are headed toward good water first. Note that heeling won't help with a winged keel.

If you remain stuck then you can use your dinghy to take an anchor to deep water and kedge off. You can use the kedge to pull from the bow, or run it from a halyard to heel the boat.

You can also get assistance in the form of a tow. Make sure the tow adds pressure gradually, and stand clear in case a line or fitting parts.

In all groundings be mindful of damage to the rudder. While the keel is designed to handle an impact, the rudder is much more vulnerable.

If the wind and waves are driving you further up, then fast action is required. Likewise, on a falling tide, speed is of the essence, while on a rising tide, patience will be rewarded.

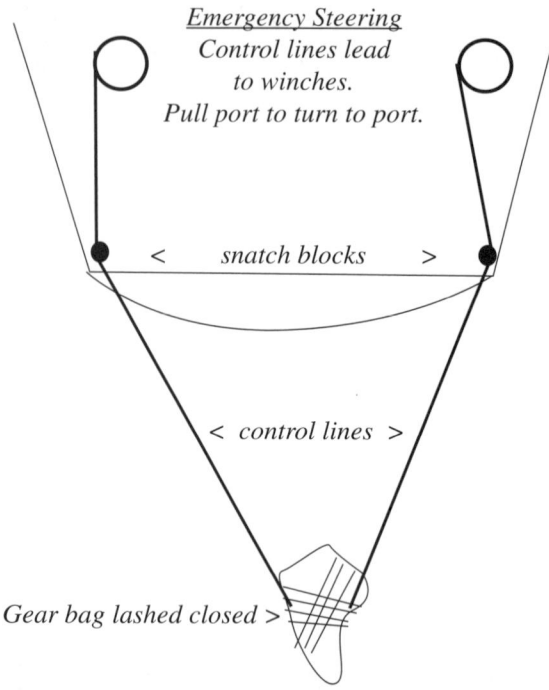

Emergency Steering
Control lines lead
to winches.
Pull port to turn to port.

< *snatch blocks* >

< *control lines* >

Gear bag lashed closed >

Dismasting

If the rig goes over the side the immediate danger is that the spar, held nearby by the rigging, will put a hole in the side of the boat. If the mast cannot be controlled, it may be necessary to cut away the rigging. For this, only hydraulic cable cutters will do. The alternative is to pull cotter pins and drive the clevis pins from the turnbuckles. For this reason, don't over turn your cotter pins, but set them so they can be removed.

Why would the spar fail? Most commonly, due to the failure of one of the many fittings in the rigging - a pin, turnbuckle, spreader or spreader base may fail. Checking all the links in this vital system must be part of your routine.

If a piece of rigging parts, but the mast doesn't tumble, then immediately turn to take load off the damaged rigging. For example, if the port spreader fails, get to starboard tack. If the forestay comes free, run off with the wind astern. Reduce sail, and jury rig temporary support for the spar with spare halyards. A spare piece of rigging and several bulldog clamps can be used to fashion a passable temporary piece of rigging. In this way you can limp to shore and effect a more complete repair.

Steering Failure

You turn the wheel, and nothing happens - or it is very hard to turn, and suddenly gets easy - too easy. Perhaps the cable has jumped the quadrant, broken, or come upclamped. The immediate challenge in close quarters is to stop the boat. Sound your horn repeatedly as a danger signal. Steer with your sails as best you can to maneuver out of harm's way, and anchor.

Often you can effect repairs on the spot. It helps to be familiar with the steering system, and how it all looked when it was working. If the problem is with the pedestal, cable, and quadrant, and it can't be fixed quickly, rig your emergency tiller to the rudder post. Obviously you'll be thrilled you've figured out how your emergency tiller sets up before you have to do it in an actual emergency.

If the rudder itself is damaged or missing then you'll have to rig up an alternative steering system. One system involves dragging objects from a bridle of lines rigged from the quarters. Pulling the load to one side turns the boat that way.

Not pictured....

Seasickness

Seasickness can be a mild inconvenience or a serious problem. Before you head out it is a good idea to review suggestions on how to prevent and deal with seasickness. Those who are prone to seasickness should take medication in advance of symptoms. Some people are happy with over the counter treatments, while others swear by ginger or wrist bands. Many find prescription Scopolamine to be effective.

The side effects of medications - from dry mouth, to dizziness or drowsiness - can be minimized by reducing dosage. Through trial and error a proper personal dosage can be determined.

There are several ways to reduce the likelihood of seasickness. The first is to adjust course and trim to make the boat's motion less upsetting. This is particularly helpful early in a cruise, until the crew get their sea legs. Also, keep crew members actively engaged in steering and trimming.

Some crew are fine on deck, but quickly grow queasy when asked to work below. Adjust your crew assignments with this in mind. Some people are set off by engine fumes or diesel odors, so be cognizant of

that. When under power or when charging batteries choose a speed and course which clear exhaust fumes.

Seasickness puts the sick crew member at risk, as lethargy and indifference, combined with an urgent need to seek the leeward rail, can lead to a fall overboard. Make sure the sick crew member is clipped to the boat. Also, change course and trim to smooth the boats motion, reduce heel, and lessen the risk.

A seasick crew member is also at risk of dehydration and exhaustion. A warm, dry bunk, sips of water, and simple crackers can help in the recovery.

In addition, seasickness puts other crew at risk, as their increased responsibilities can lead to fatigue, and with it, errors.

A further problem is the spread of seasickness, as the mal-de-mer of one increases the likelihood of trouble for others.

As with all hazards, prevention is preferred to treatment. No sailor is immune - so take precautions by changing course if the boat's motion is unsettling, and by watching over each crew member for the first signs of trouble.

Illness and injury

Fatigue, overworked muscles, a pitching, heeling wet slippery deck. The challenges of moving about on deck and below make bumps and bruises routine. The chance for serious injury is ever present.

On a long cruise some of the crew should be trained in first aid. You should gather your crew's medical profile, and prepare a medical kit with the help of your doctor.

The greatest danger is from head injuries due to a swinging boom, and from drowning after falling overboard.

The most common injuries are sprains and bruises. Broken hands and fingers are also not uncommon. By way of prevention, train your crew not to fend off if your boat is going to run into something. Boats are easier to fix than hands.

Other common problems include dehydration due to heat or seasickness, constipation, sunburn, and hypothermia. So, drink plenty of water, eat fruit, and protect yourself from the elements.

Risks and Rewards

With all the safety issues and risks associated with cruising, it may seem a wonder anyone survives. Safe and successful cruising requires a high degree of self sufficiency. At the same time, while at sea, we are less subject to the hazards imposed on us by others ashore.

Prepare, prevent, remain forever vigilant, and remember - the only way to escape the risks of cruising is to face the ever-present risks of life ashore.

7. Equipment List

Following is a consolidated list of equipment and preparations for Coastal and Offshore Cruising:

Below Decks:
• Heavy movable items such as batteries, stoves, gas bottles, tanks, tool boxes and anchors and chain should be securely fastened.
• Soft wood plugs, tapered and of the appropriate size, should be attached or stowed adjacent to every through-hull opening.
• Bilge pump handles should be permanently installed or secured by a lanyard.
• Bunks should be fitted with lee cloths or some other means to allow rest when the boat is heeled or pitching.
• The mast should be fastened to the mast step to prevent the mast from jumping the step in the event of a rigging failure.
• The engine should have a dedicated starter battery, isolated from the ships batteries.
• Each fuel tank (for engines and stoves) must have a shut-off valve.
• Companionway hatch boards must be secured to stay in place when inverted, with the hatch open or closed, and be operable from above and below deck. When not in place hatch boards must be secured to the boat (e.g. by lanyard) to prevent their being lost overboard.
• Fire extinguishers accessible from different parts of the boat - i.e., on deck, forward, and near the galley. The engine box should have a fire extinguisher access port.
• Sufficient hand holds to allow safe movement about the boat – both on deck and below.
• Put nonskid on hatch covers, and ladder steps.

On Deck:

- Lifeline gates should be taped shut during a passage, and the pins and rings at the terminals should be taped over as well.
- Jacklines to allow movement the length of the boat on either side of the mast without unclipping.
- Secure fittings to allow clipping on before coming on deck and after going below, and while working at the helm, sheet, and halyard stations, and at the mast base.

Electronics and Navigation:

- Running lights must be visible when under sail. A mast head tricolor has far greater range than pulpit mounted lights.
- VHF with masthead antenna, plus an emergency antenna. A handheld VHF, in addition to a permanently mounted VHF, is suggested.
- GPS with chart system. A handheld GPS, in addition to a permanently mounted GPS, is suggested.
- Spotlight – either running from ships power, or self powered, for use in finding navigation aids or crew overboard.
- Foghorn
- Bell
- Radar reflector
- Charts, light lists, and plotting equipment.
- Depth sounder or lead line.
- Knotmeter
- Compasses

Other Equipment:

- At least 2 anchors, with chain and rope, ready for immediate use. One may be stowed below, or in a locker aft, but be careful not to bury it.
- Emergency steering:
  - emergency tiller in the event of wheel or quadrant failure.
  - alternative means of steering in the event of rudder loss.
- Cutting tools to sever rigging in the event of mast failure.

Properly sized soft wood plugs should be secured on a lanyard near each through hull fitting.

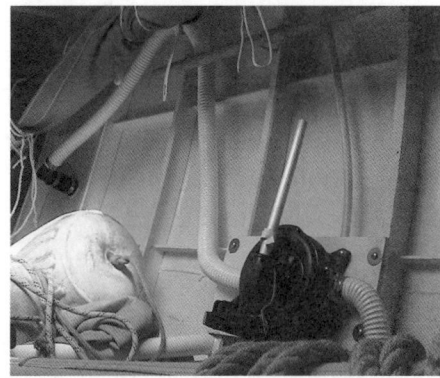

Bilge pump handles should be permanently fitted, or secured by a lanyard. Note the high loop in this hose to prevent siphoning when the boat is heeled.

Emergency calling information should be posted adjacent to the radio. Also, show your crew the Crew Overboard button on your GPS (or list the button sequence to record and then go to the current location.) Note the fire extinguisher mounted under this navigator's table.

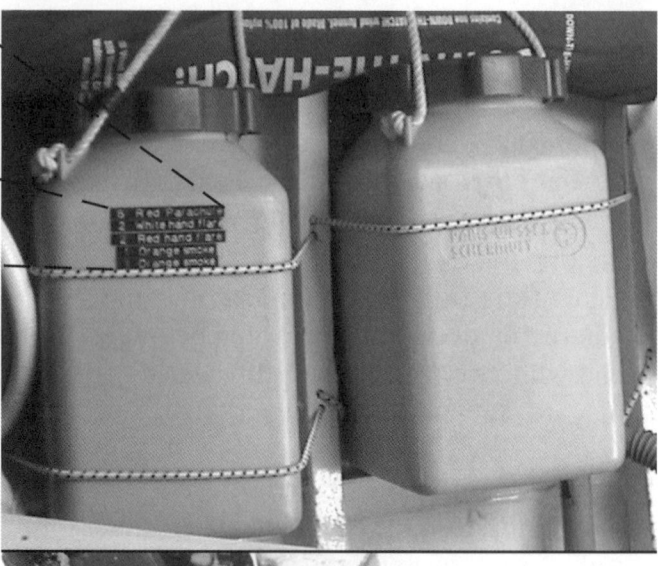

> 6 Red Parachute
> 2 White hand flare
> 2 Red hand flare
> 1 Orange smoke
> 1 Orange smoke

SOLAS Flairs are dramatically more effective than those which meet the minimum US Coast Guard requirements. They must be stowed in a water tight container, and kept up to date.

Personal Gear:
- Lifejacket with whistle and reflective tape.
- Safety harness and tether.
- Flashlights - at least one per crew member, plus spare bulbs and batteries.
- First aid kit and manual, and first aid training.

Emergency Equipment:
- Heavy weather jib and storm jib with lead positions marked on deck.
- Storm trysail rigged separately from the mainsail.
- Yacht's name and retroreflective tape on all floating gear. (i.e. lifejackets, cushions.)
- LifeSling with reflective tape and light
- Lifebuoy or Overboard Module with reflective tape and light
- Heaving line - throwing sock type.
- Flares - SOLAS type. 4 each of: Red parachute, Red hand held, Orange smoke and White smoke.

For Offshore Passages:
- Liferaft
  - with canopy and insulated floor
  - stowed where it can be accessed in 15 seconds or less.
- Grab bag for liferaft including some items already listed. A *grab bag* is a way to keep these items organized and together in the event you need to make a quick getaway. Items include:

Flares which have not been properly stored are less than useless.

- waterproof handheld VHF
- waterproof handheld GPS
- EPIRB - GPRIRB 406 type, which provide yacht name and location when activated.
- a first aid kit and seasickness tablets
- two or more *Cyalume* glo-sticks
- two watertight floating flashlights
- one daylight signalling mirror
- one signalling whistle
- additional flares
- rations
- water (at least one pint per person)
- prescription glasses and medication

Chapter VI –
Knots, Line and Gear

1. Rope and Line

2. Knots

3. Lines Under Load

Knots, Line and Gear

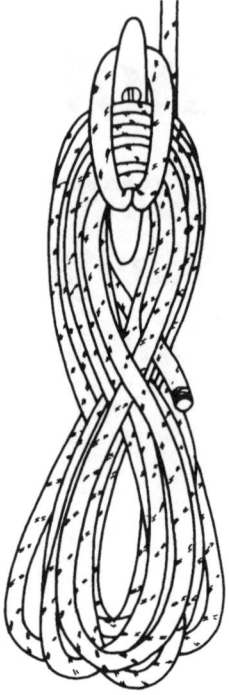

Coils of braided line should be allowed to fall into figure 8s. Coiling in loops causes kinks. Only twisted three strand line should be coiled in loops.

1. Line and Rope

Rope is the stuff that, once it's cut up and put on a boat, is called line. Lines include halyards, sheets, anchor rodes, outhauls – you name it.

Different types of rope have different characteristics that make them good for specific uses. Some rope is laid in distinct strands, while other rope is braided. Laid rope usually is a bit stretchier.

Halyards and sheets should not stretch, so they're made of low-stretch Dacron or space-age aramid fibers. Halyards on older boats may be made of wire rope, but most boats today use regular low-stretch rope, which is easier to handle and just as strong, if not stronger.

Anchor rodes and docking lines should stretch to absorb shock loads, and that's why they're made of nylon.

Coiling

When coiling line, loop it onto one hand. If it's laid rope (3 strand), twist your right wrist so it hangs in even loops. If it's braided, don't twist your wrist; it should hang in a figure-8 (otherwise it will kink).

Heaving a Line

A fine art when docking or picking someone out of the water is heaving a line accurately.

Throw three coils while letting the rest run off your non-throwing hand. Aim high, and upwind of your target. Practice with lines of different size and length.

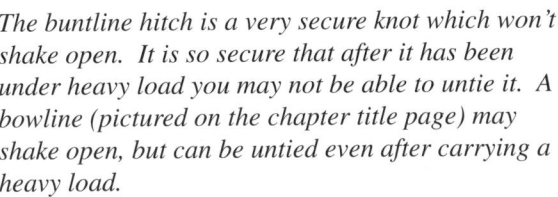

The buntline hitch is a very secure knot which won't shake open. It is so secure that after it has been under heavy load you may not be able to untie it. A bowline (pictured on the chapter title page) may shake open, but can be untied even after carrying a heavy load.

2. Knots

There are plenty of knots, and learning their uses and how to tie them is fun. Here we'll present a few knots we feel are essential.

When choosing a knot, consider its purpose and also two factors: How easy is it to untie? How secure is it?

The Stopper Knot

The stopper knot used to keep a line from running out can be an overhand knot or a figure-8. Both are simple to tie, but the overhand can be hard to untie if it's pulled hard, which means it's more secure. Therefore the overhand is the stopper knot of choice for the end of a halyard (which is rarely pulled out), but the figure-8 is best for the end of a sheet that will have to be untied and put away each afternoon.

Here's a tip: when tying any knot, always leave at least several inches between the knot and the bitter end so when the knot slips under strain (and it will), it won't come undone. Also, if the knot runs up against a cleat or block, you can still grip the end.

The Bowline and Buntline Hitch

Almost all knots on a sailboat are tied to form loops.

The bowline is the most valuable knot. You should be able to tie one blindfolded and quickly. Because it's easy to tie and untie, and also pretty secure, it's a good knot for tying jib sheets to the clew of the jib. You will want to be able to remove the sheets.

The problem with the bowline is also its value: it's easy to untie, which means that it can untie itself when not under strain. If it's on deck, you can always check it, but a loop in a remote place, like the top of the mast, requires a more permanent knot.

When tying a halyard to a shackle or a reefing line to the boom, use the buntline hitch. While it may have to be cut in order to be removed, it won't shake itself loose. It looks complicated but actually is pretty simple.

The Clove Hitch and Double Half-Hitch

Less permanent, more easily tied loops are sometimes called for.

To secure a fender to a lifeline, try the clove hitch and its cousin, the double half-hitch. The advantage of these knots is that the length is more easily adjusted.

The trouble is, they can slip open, so be careful. Taking an extra turn around the object makes the knot a little more secure.

While these knots are easier to tie than bowlines, use the good, secure bowline for important jobs, like securing docking lines and the dinghy painter.

Securing the Anchor Rode

There must be no question that the anchor rode is secure.

An excellent knot here is the fisherman's bend, also called the anchor bend. Note how the line is locked within the knot.

When tying the rope rode to the anchor or chain, secure the bitter end by either tucking it into a strand or seizing it with light line. Better still, use an eye splice on a thimble. The shackle pin should also be moused shut with wire or light line.

Tying Lines Together

Occasionally the ends of lines must be tied together. If there is enough line, the preferred way to secure two lines is with two bowlines. Another excellent choice is the double figure 8. Tie a figure 8 loosely in one line, and retrace it "upstream" with the second line, so the free ends lie opposite, not together.

The sheet bend is used to secure two lines of different size to each other. Form a loop with the thicker line, with the tail on top of the standing part. As with a bowline,

The clove hitch, at left, is good for hanging fenders, as the length is easy to adjust. Once set, add a half hitch to prevent slipping. The cow hitch, at right, should be used to tie up cows. It slips easily and is the culprit in many cases of lost fenders.

A splice is preferred, but if you must tie an anchor rode, use an anchor bend. Note the tuck in the end of the line, and also the seizing on the shackle.

pass the smaller line up through the loop, around the standing part, and back down through the loop.

The square knot can jam badly under load, or trip if loaded out of line. Don't count on it. But if one end is looped through, you have an easily untied reef knot.

All knots reduce line strength by as much as 50%. For permanent solutions an eye splice and thimble is preferred, as it preserves more of the line strength.

To reduce wear and prevent creep, keep working halyards secured to a winch as long as you have a winch available. Use enough wraps that the line is easy to hold coming off the drum.

3. Lines Under Load

A sheet or halyard that's under load can pull very hard, so treat it with respect. Always snub a loaded line around a cleat or winch. Keep your hair, jewelry, and fingers well clear of cleats and winches.

If your halyards are rigged through rope clutches, it is still best to keep any active halyards cleated to a winch. This controls 'creep' or slippage, keeps the halyard set up for adjustment, and increases the life of the halyard, as it is not spot loaded.

To cleat a line on a traditional horn cleat, take one full turn around the base, then make at least one figure-8 over the cleat's horns, and finally lock it off with a half-hitch. If you find that the half-hitch is hard to undo, you're not making enough figure-8s.

When trimming a line on a winch, start with two wraps and add two or three more wraps as the load builds. Too many wraps at the start will cause the line to override, and jam in itself. When adding wraps to a winch, use one hand, keep an overhand grip on the line, and keep your finger tips clear.

One way to control the line is to press it against the winch drum with the palm of your hand. To ease a few inches hold the free end in your left hand, place your right palm on the drum, and push the line out.

A coiled halyard can hang from the halyard winch. Note that braided line coils into figure 8s - not loops.

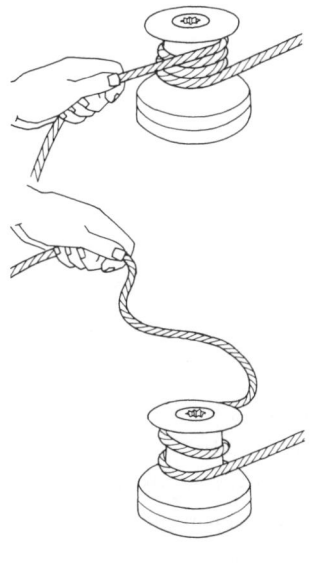

To cast off a line, ease an armload, then lift the line straight off the top of the winch and let it run out through your hands.

To cast off the line, ease an armload, then lift the line straight off the top of the winch and let it run out through your hands.

A coiled halyard can be laid over the winch. Before dropping the sail fake the halyard out on deck to be sure it will run without kinks.

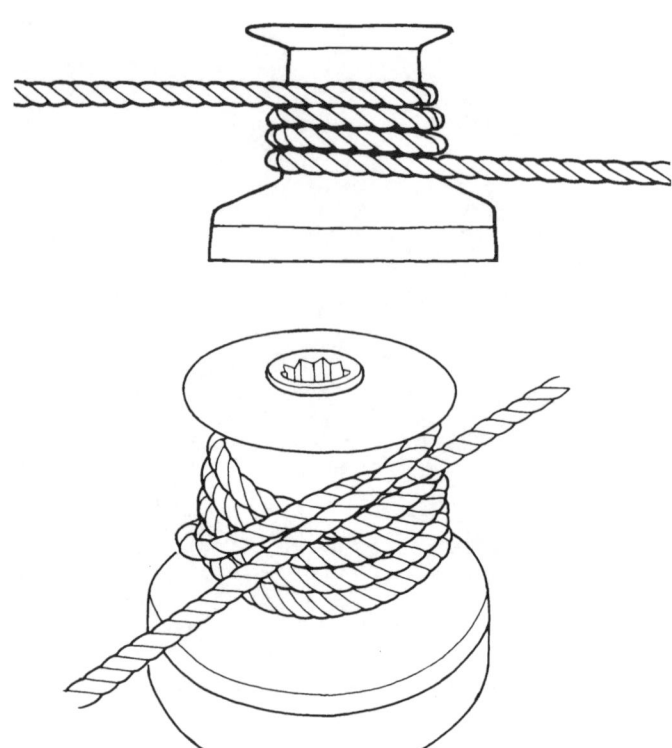

Overrides result from slack in the line and too many wraps on the drum when trimming or easing. Clear overrides by tieing in a short sheet to take the load.

To prevent roller furler overrides keep tension on the furling line when unrolling the genoa.

Winch Overrides

If there are more than two turns of a sheet, halyard, or other line on a winch before the load comes on, one turn may very likely ride over another and jam the line. Overrides are also created when easing a line on a winch loaded with excess wraps.

To remove the override, take the load off the line forward of or above the winch and straighten out the turns. If it's a jib sheet, rig and tension another sheet to a winch. If it's a halyard, take the load above the winch. Or using a spare line (called a "short sheet"), tie a rolling hitch to the fouled line forward of or above the winch, then pull on the short sheet.

You can often take the load off a jib sheet by heading up and luffing the sail, or by lowering the jib halyard a few feet.

Overrides on the roller furling drum can be difficult to clear. To prevent them keep tension on the furling line as the sail unrolls. Beware, as the filling jib can pull hard and fast on the furling line, and cause a rope burn.

To clear the override you'll need to slack the furling line and clear the line by hand. No mean feat, particularly since the task is performed underway. Careful use and prevention are suggested.

Chapter VII –
Anchoring and Docking

1. Anchoring
2. Docking

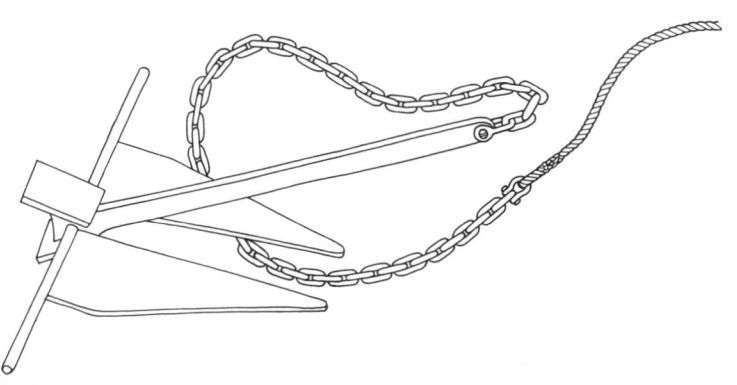

Anchoring and Docking

For cruising it is best to carry (at least) two anchors, with one ready for immediate use. A CQR is shown at left; a lightweight Danforth- type above. Note the chain lead for the rode. Shackles on the anchor rode should be wrenched tight and seized shut with wire or wire ties. Another popular anchor is the Bruce anchor, below.

1. Anchoring

The Anchor and Rode

Anchor ground tackle has two parts. The anchor and the rode – the rope and chain (or just chain) that connects the anchor to the boat. Different anchors work best in different bottoms. The most secure bottom is soft mud or sand, but you may have to deal with hard mud, rock, or gravel.

Burying anchors with sharp-pointed flukes dig into a soft bottom and can be relatively lightweight. The most popular burying anchors are the lightweight Danforth-type, the plow or CQR ("secure" — get it?), and the Bruce.

Non-burying anchors, like the traditional yachtsman's anchor, often work best on hard bottoms, where their great weight provides most of the needed holding power.

A boat's pull on an anchor is a function of her resistance to the wind and waves — which means windage (beam and rig) and displacement. Anchor manufacturers provide recommended sizes and weights for different boats. When in doubt, go for the bigger anchor.

A rode must be strong and also able to absorb shock loads when the boat pulls and swings in gusts or waves. Otherwise the anchor will trip, or break loose. There are two types of rode: rope and chain. A rope rode should have a chain lead (pronounced "leed") to minimize chafe on the bottom. A rope rode must be nylon – not Dacron – because it stretches and absorbs shocks.

A rope rode with a chain lead is the most common choice because an all-chain rode is

By matching your scope and boat type to those around you, you assure that all boats will swing in a similar manner as wind and current change.

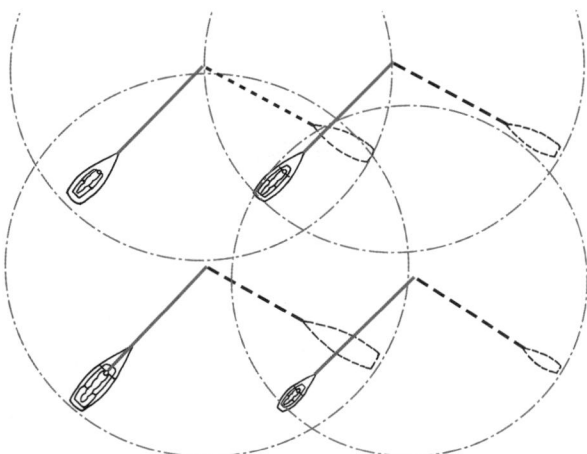

very heavy, hard on the hands, and requires a special windlass (a winch for handling ground tackle). The lead can be as short as 6 feet; though 10-20 feet is preferred.

Chain resists chafe (wear) better than rope and absorbs shocks by lifting its heavy weight off the bottom. An all-chain rode is common among long-distance cruisers and people who often anchor on rocky bottoms.

Mark the rode every fathom (6 feet) or every 10 feet so you know how much you've let out.

Choosing the Anchorage

The anchorage should be out of traffic and swift tides, and protected from the prevailing wind and waves. The water should be plenty deep for your boat, and not so crowded that you can't use proper scope. And the bottom should be suitable for your anchor. Charts show the type of bottom ("sft M" is soft mud, and "rky" is rocky) .

Generally speaking, if boats like yours are anchored there, it's probably safe for you – if you keep a careful lookout.

By anchoring near boats like yours, and using similar scope, you can assure that you will swing in a similar manner as wind and current change.

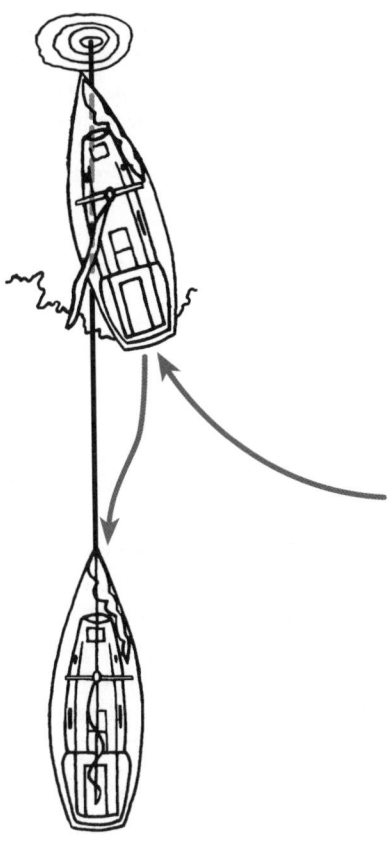

Turn into the wind, and lower the anchor over the bow as the boat stops. Back away and ease out scope to 7 times the depth plus freeboard, then back down hard to dig the anchor in and to make sure it is set.

Anchoring

The steps are straight forward, but require full attention. It helps to arrange hand signals so the crew on the foredeck and the steerer aft can communicate with each other, particularly if the engine is running.

- Flake the rode so it will run out easily, and *always* tie off the bitter end to the boat.
- Power or sail into the wind (or current – whichever is stronger) until the boat is stopped where you want to anchor.
- Lower the anchor over the bow (don't throw it!) and back down, either by motoring in reverse or backing your sails. Veer out rode until the *scope* is about 7:1.
- Cleat the rode and back down hard to dig in the anchor. When the bow dips and a heavy load comes on the rode, you are

"Scope" is the ratio of the length of rode deployed to the height from the bottom to the deck (depth plus freeboard). Scope of 5:1 is generally adequate, while 7:1, or more, provides better holding in strong winds or waves.

Scope Ratio	Anchor Angle (a)
2	55
3	69
4	75
5	78
7	82
10	84
20	87

As the table suggests, increasing scope provides diminishing returns in terms of the angle at which the rode pulls the anchor. Additional scope does provide more shock absorber to protect the anchor from shock loads. On the downside, extra scope also gives the boat more room to swing, and increases the propensity to "sail" at anchor.

dug in. Touch the rope. If it jerks or wobbles, you're dragging and you'll have to haul up the anchor and try again, perhaps in another place. The anchor may have failed due to weed or rock, or it may be clogged with mud. If the rode is firm, then the anchor is set.

• In a crowded anchorage match your scope to the boats around you by matching the angle at which your rode hits the water.

• Once you're anchored, locate a range or take bearings ashore that tell you if you're dragging. Also be alert to the tide table, since a change of current may trip the anchor, and a rising tide reduces your scope. Select items which will be visible after dark so you can check your bearings through the night.

Scope

An essential concept in anchoring is *scope*. Scope is the ratio between the amount of rode let out and the depth from the boat's deck to the bottom. The greater the scope, the narrower the angle between the rode and the bottom.

The rode must lie at a narrow angle to the bottom so it pulls the anchor back (digging it in) and not up (tripping it or pulling it out).

Nylon rode is nearly weightless under water, but chain's weight causes sag in a catenary, and that keeps the rode near the bottom. But any rode can be made to lie low by increasing its scope. Remember that the main factors are the boat's windage (beam and masts) and displacement.

How much scope should there be? Here are some rules of thumb:

• In moderate wind and smooth water, the scope on a boat of moderate displacement and beam should be at least 5:1 (50 feet of rode for every 10 feet from the deck to the water bottom). On an all-chain rode the scope should be at least 3:1 in moderate wind.

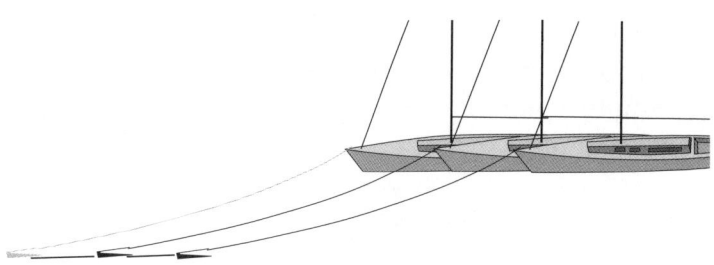

If your anchor drags, then lift and clear it, and try to reset with more scope. If that fails, search out better holding or a more protected spot, or set a second anchor at an angle to the first.

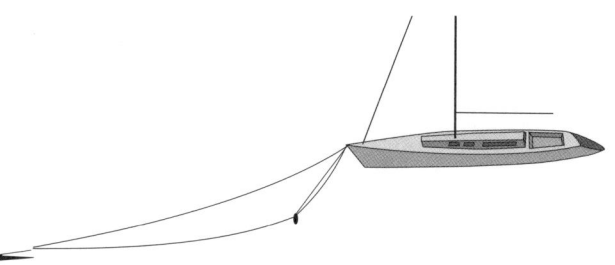

A weighted sentinel, also called a kellet, can improve anchor performance by improving the rode's angle to the bottom and by absorbing shock loads.

- As the wind builds, increase scope. In a strong wind, a rope rode should have scope of at least 7:1. For chain a scope of at least 5:1 is recommended.
- As long as there is room to swing, it is hard to have too much scope. The load angle changes little as scope increases from 7:1 to 10:1 or more. The advantage of extra scope is the ability to absorb shock and protect the anchor from being jerked out of the bottom.
- If room does not allow sufficient scope, then a weighted sentinel can be added to the rode to improve the angle to the bottom and to absorb shock loads. A float further up the rode can provide additional shock load protection.
- Due to increased windage, multihulls and boats with high freeboard need proportionately larger anchors than conventional monohulls.

Dragging Anchor

Even with plenty of scope the anchor can drag if the holding ground is not good, or the anchor is not big enough, or sufficiently dug in, to handle the load on the rode.

Most GPS devices have "anchor watch" alarms that signal if the boat moves, but you may not need it. You may be so attuned to changes in wind, waves and your boat's motion that you'll wake to any change. If something feels different, then check your bearings.

To tell if the anchor is, in fact, dragging go forward and put a hand on the anchor rode. A firm steady pull shows an anchor that is holding. A wobbling or shaky feel is a dragging anchor.

If your anchor does drag, pay out more scope. Failing that, raise and clear it before you try to reset. If it still won't hold, then move to another location with a different bottom.

In building wind and chop it may be difficult to get the anchor to dig in. Light weight anchors may sail or skip across the bottom if the boat drifts back too fast. It can help to run the engine ahead to slow your drift and allow the anchor to dig in.

If another boat's anchor drags, sound your horn to get the crew's attention. Prepare to fend off.

Set Two

While extra scope, chain, and a sentinel can improve holding power, nothing improves holding power like a second anchor.

There are a number of occasions when setting two anchors is a good idea.

In rough weather set a second anchor at 45 to 60 degrees to the first anchor. Setting two anchors divides the load, and reduces the tendency of the boat to *sail*, or swing on its anchor. Hanging a weight, or sentinel off the anchor rode can improve anchor performance, and also reduce the chances of fouling on the keel or rudder, particularly when lying to two anchors.

Another time to consider a second anchor is when you are anchored in an area of strong tidal currents. Set one anchor to the current current, and set the second to hold when the current reverses at the change of the tide.

Two anchors are also called for when you anchor near to and parallel to shore, shoals, or reefs. If the scope of your primary anchor is such that you would swing into shallows in a wind shift, then set a second anchor to hold you off. Use a sentinel to hold the second rode down so it won't be a hazard to passing traffic.

Before you run off to have a second bow roller added to your stem fitting, understand that setting a second anchor creates its own set of problems. For starters, in an anchorage, your boat will swing differently than those lying to a single hook, which could result in a tangle or collision. Also, as mentioned above, the second rode can tangle on the keel, prop, or rudder, or snag the first anchor or tangle in the rode.

Set Three and Leave

If your boat is threatened by a tropical storm or hurricane, then strip the sails and any excess gear from the deck, set all your anchors, double up chafing gear, and get off the boat. Do not put your life at risk to try to protect your boat. Besides, there is little you can do at the height of a storm. Get off.

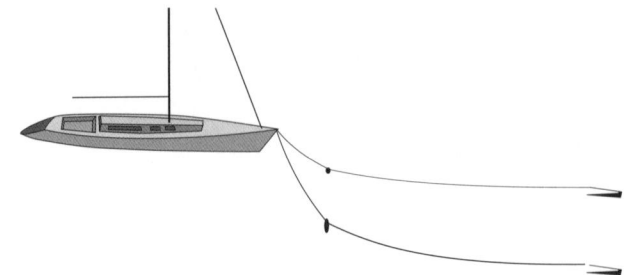

Set a second anchor to provide extra holding and to reduce 'sailing' at anchor.

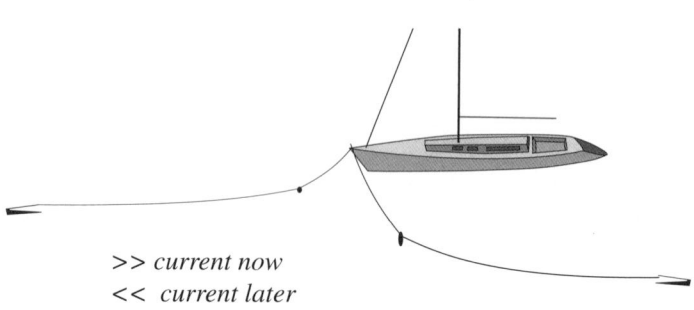

>> *current now*
<< *current later*

When anchored in a strong tidal current it is also worthwhile to set a second anchor in anticipation of the current reversing as the tide changes.

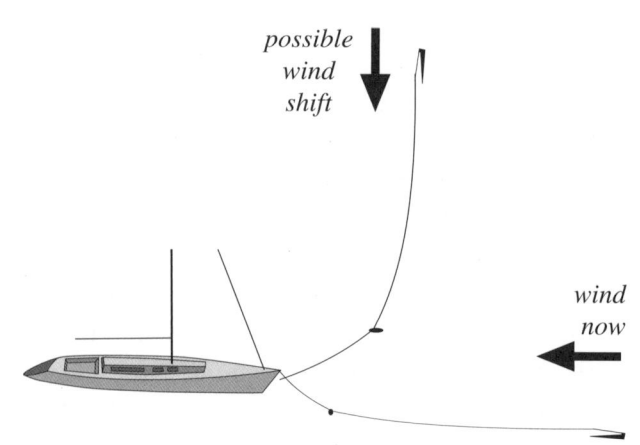

possible wind shift

wind now

shoals

Set a second anchor to keep you off a nearby shoal or shoreline in the event of a windshift.

To weigh anchor take up slack until the anchor rode is "straight up and down," and then pull the rode, or motor ahead slowly to break the anchor out. Then haul away....

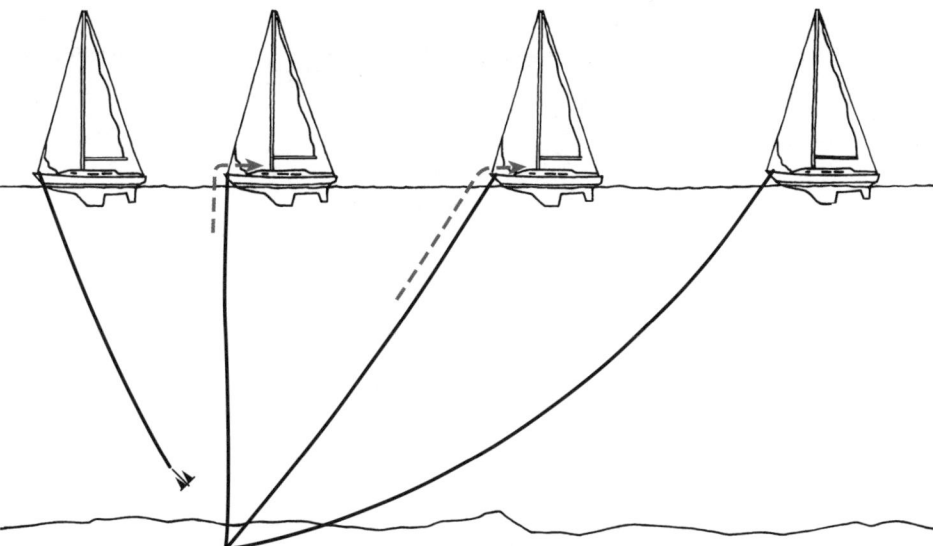

Weighing Anchor

Weighing (hoisting) the anchor is like anchoring, only in reverse.

Pull forward under power or sail until the bow is over the anchor ("straight up and down"), cleat the rode, then move ahead some more to break the anchor out. Stop the boat while you haul to keep the rode from chafing the topsides.

You need not motor to weigh anchor. You can sail up to your anchor by short tacking up to it; you may have to back the jib and pay out some scope initially. This is a challenging but pleasant way to get underway without running the engine and waking the entire crew.

When the rode loosens, haul up the anchor - *many hands make light work, and electric windlasses are great too*- clean it, and stow it.

A Fouled Anchor

If the anchor won't break out try pulling ahead, or circling around to pull from a different direction. It may be tangled under another rode, or caught on a rock. When anchoring in rocky areas rig a trip line off the crown of the anchor. Rig a buoy to float

When anchoring among rocks rig a trip line and buoy off the anchor crown. Pull the trip line to back the anchor out if it is wedged into rocks.

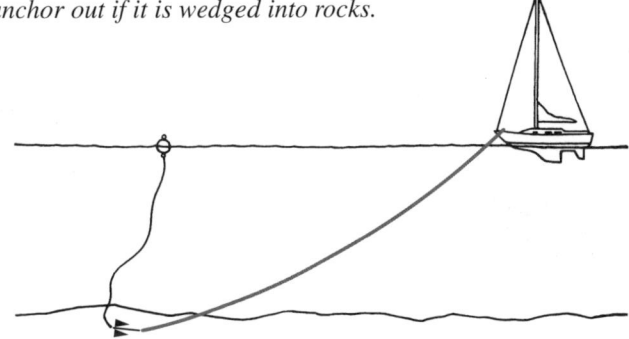

it above the anchor. You can then use the trip line to back the anchor out.

Another way to get the anchor unstuck is to take up as much slack as possible with the boat directly over the anchor, and let the buoyancy of the boat lift the anchor as the boat bobs in waves, taking up slack as it becomes available. It is a more passive approach than powering against the rode, but it works.

Another option if it is not too deep is to dive down the rode and attempt to clear the anchor by hand. This can be a tough task if you don't have full dive gear.

If all else fails, mark the rode with a float and come back later to try again...

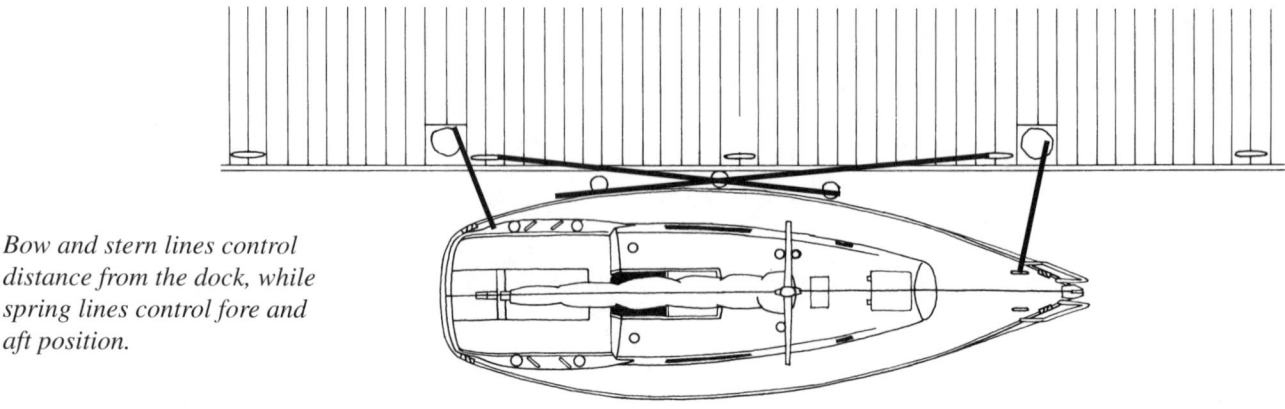

Bow and stern lines control distance from the dock, while spring lines control fore and aft position.

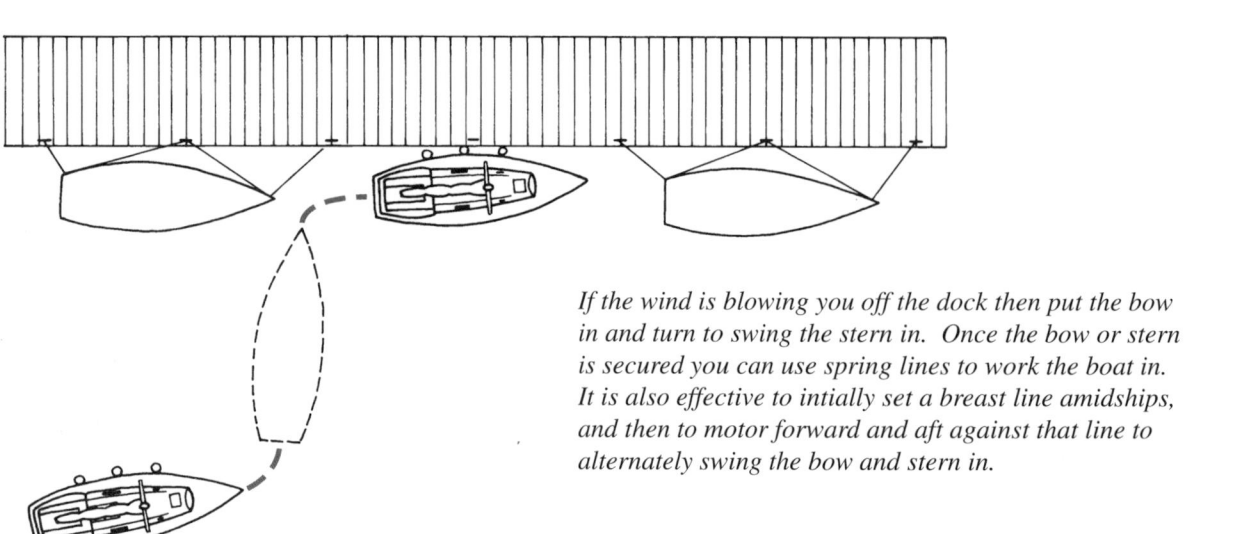

If the wind is blowing you off the dock then put the bow in and turn to swing the stern in. Once the bow or stern is secured you can use spring lines to work the boat in. It is also effective to intially set a breast line amidships, and then to motor forward and aft against that line to alternately swing the bow and stern in.

2. Docking

The keys to hassle-free docking are using docking lines properly and knowing how your boat behaves under power at low speeds. Practice with plenty of fenders.

The four docking lines are the bow line, the stern line, and the two spring lines. The bow and stern lines control distance from the dock. Spring lines are intermediate lines in an X-arrangement that control the boats fore and aft movement. The forward spring runs forward from the boat to the dock, and the after spring runs from the boat (you guessed it!) aft to the dock.

You can use dock lines to advantage when undocking and docking.

When undocking, to swing the bow out and the stern in, power in reverse against the forward spring or stern line. To swing the stern out, power forward against the after spring line.

To avoid having to leap off the dock, double your dock lines back to the boat. That is, the dock lines should run from the boat, around a cleat or piling on the dock, and then back to the boat, so you handle the lines from onboard.

When docking, come in as parallel to the float as you can. First set the after spring and stern lines. Next, as the boat moves ahead, and the bow swings in, set and secure the bow line and forward spring. You can also set the stern line first, and motor ahead gently to bring the bow in.

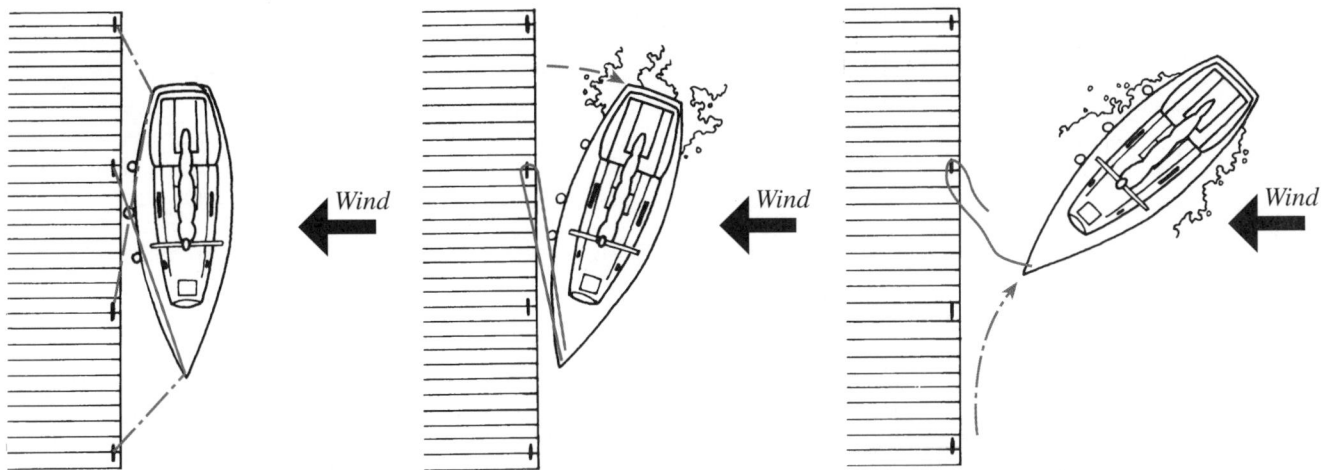

Top: With the breeze blowing the boat against the dock, set fenders forward and kick the stern out by motoring ahead against the aft spring, then back out hard.
Bottom: With the breeze on the bow, back down against the forward spring to kick the bow out, and then motor ahead.

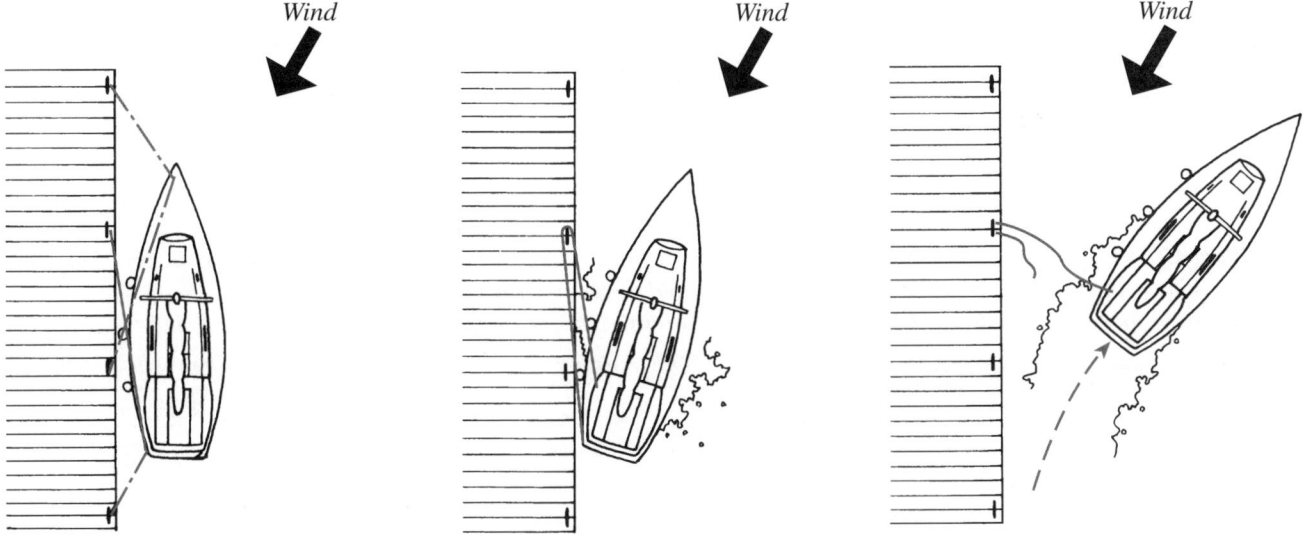

Avoid tossing your bow line to a helpful volunteer on the dock. If your helper pulls the bow line tight it can smash the bow into the dock. It is better to toss a stern line.

For shorthanded docking rig an extra-long dock line from bow to stern, and step off the boat with that one line. Secure the boat, and then set your regular dock lines.

When the breeze is parallel to the dock, approach with your bow into the wind so the breeze will help slow you.

If the breeze is holding you off the dock, then put the bow in, and set the bow line and forward spring. Swing the stern in by backing down against the spring line. A standard (right hand) propeller "walks" the stern to starboard in forward gear, and to port in reverse. If you come in port side to, then kicking the engine into reverse will walk the stern in.

When the wind is blowing in, pull up nearly parallel to the dock, but with the bow slightly out, as it will blow down faster. You may want to set up port side out in anticipation of backing out against the breeze.

In anticipation of a big storm double up your dock lines and chafing gear. If you'll be leaving your boat for a period of time, do the same. And leave a spare dockline coiled in the cockpit to make it easy for a good samaritan to do a good deed.

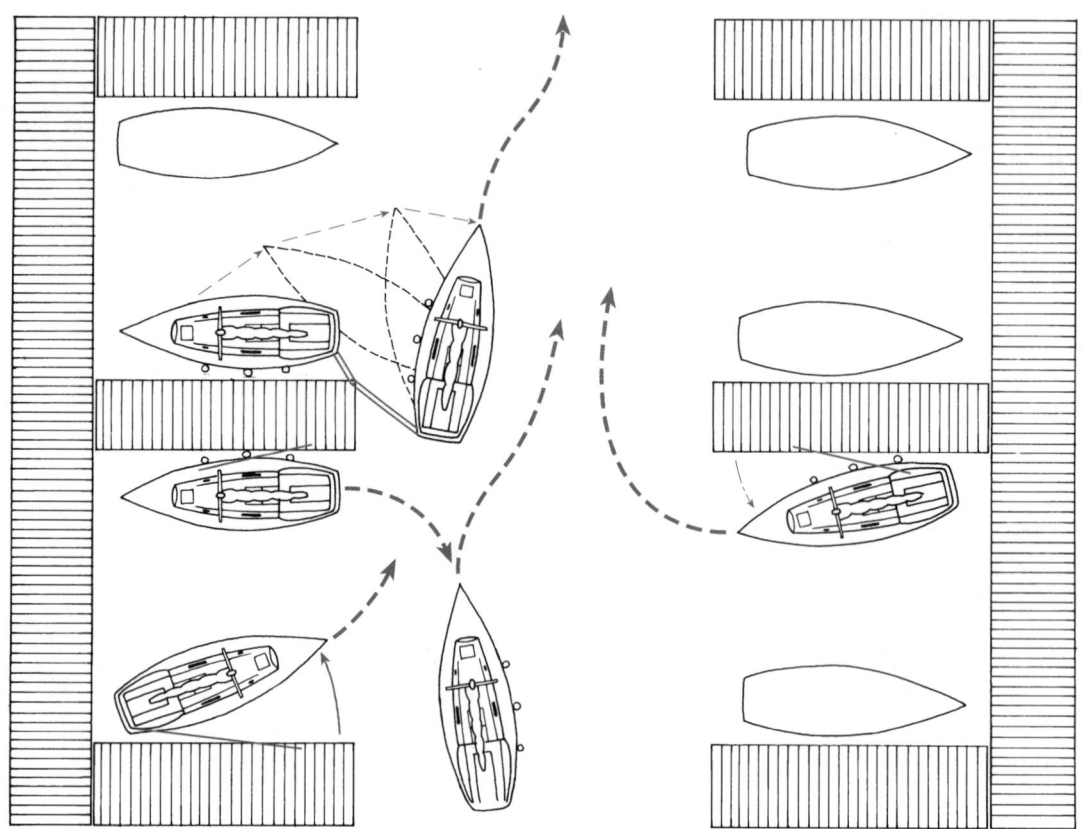

Top: When departing a slip, use your spring lines to keep the boat from blowing down, and to help turn the bow out.
Bottom: When coming in to a tight slip, get lines ashore, and then work your way in.
Alternately, if the fairway is downwind, it is sometimes easier to back down the fairway, and even back into the slip.

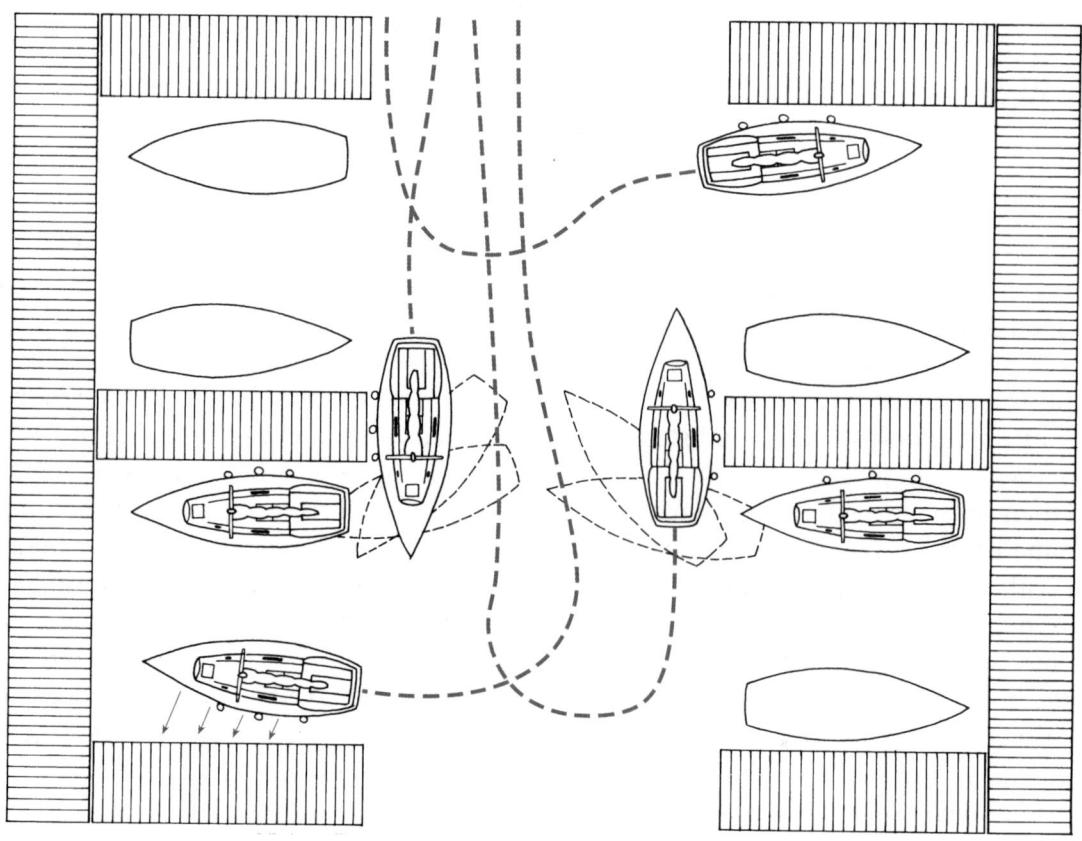

Chapter VIII –
Sailing to a Destination

1. Welcome Aboard
2. Dead Reckoning, Coastal Piloting, Navigation Aids, Tides, Current, and Traffic
3. Sailing in Wind Shifts
4. Sailing at Night
5. Problems Along the Way
6. Conclusion

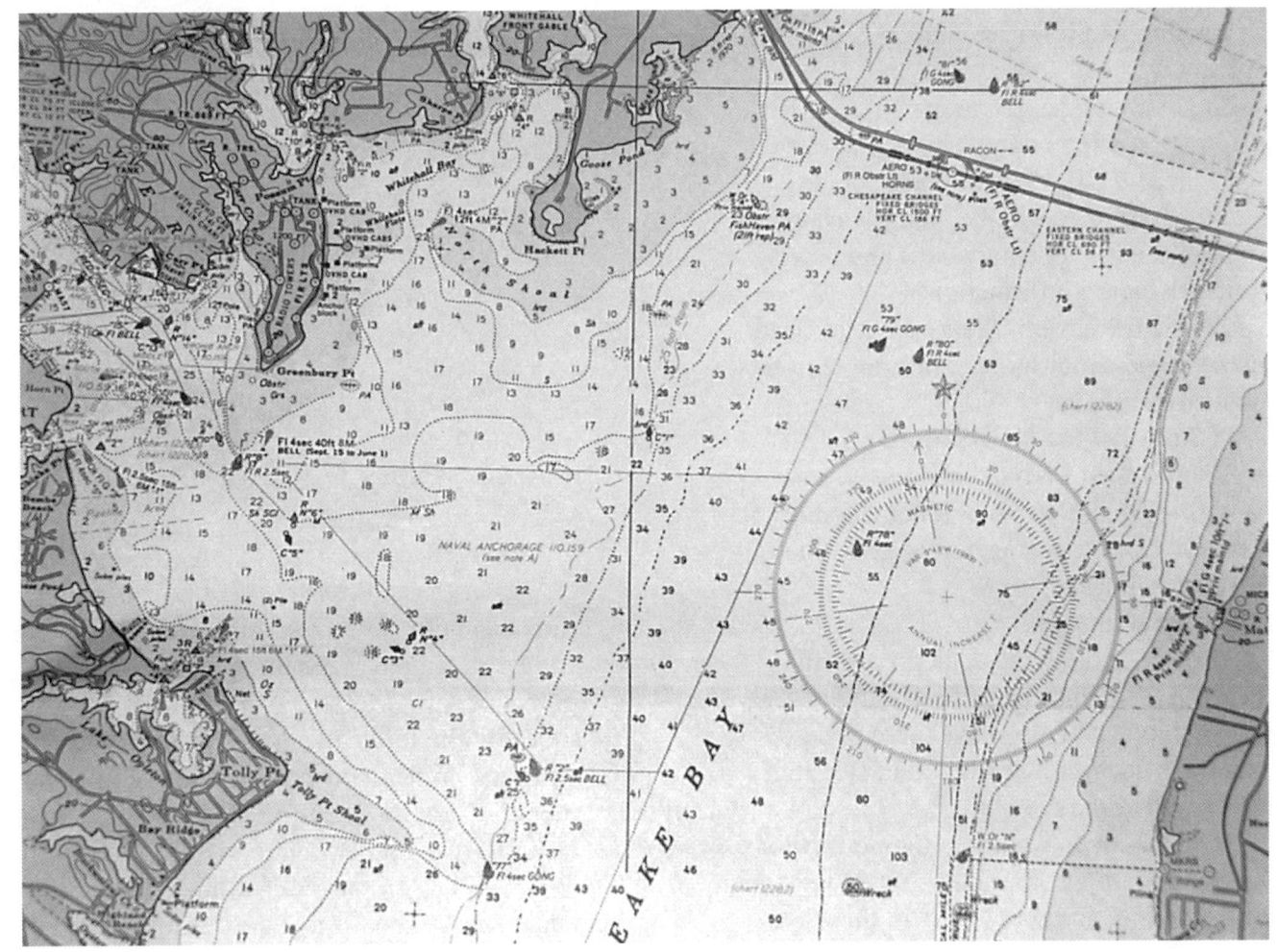

Sailing to a Destination

1. Welcome Aboard

A skipper welcoming guests on board should make every effort to look at the situation from that person's point of view. What would that person want to know right away? What else do they need to know? The details you cover will, of course, depend on the experience level of the new crew member. Here's a check list of items you might run through with guests who sail:

Check list for sailing guests
- Life jackets to each crew member
- Safe winch use with lines under load
- Sequence of getting under way:
  Leaving the dock (under power)
  Rigging and hoisting
  Trimming technique
  Tacking and jibing technique
- How the head works
- Fire extinguisher locations
- Location of an emergency call placard near the VHF. (This placard should explain how to use the radio, what channel to call in an emergency, and how to find location information.)

There are several things you can do to make a non-sailor more comfortable in this new environment.

Check list for Non-Sailors
- What to do if you feel queasy
- Will the boat tip over? It helps to heel at the dock or mooring before you head out.
- What to hold on to
- What not to touch

Before leaving the dock go over the expected sequence of getting underway, as listed above, and also have them practice handling lines and winches, and steering, while still at the dock.

Once in open water put the new sailor to work - trimming, steering - so they feel part of the action, and don't feel "in the way."

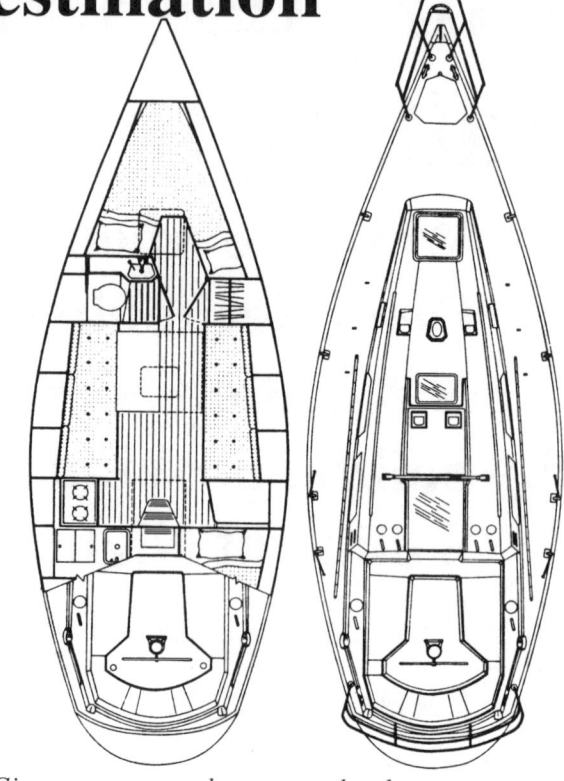

Give new crew members a complete boat tour. To help new crew get accustomed to a world which "leans," heel the boat using crew weight before you even leave the dock.

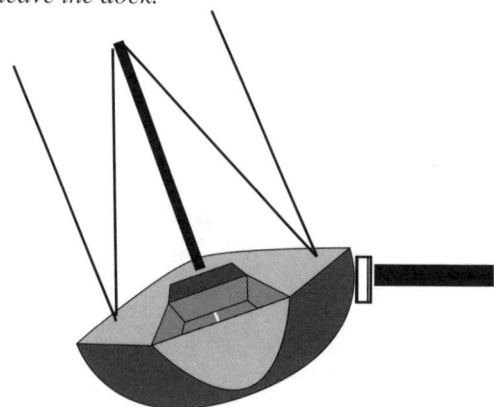

A very brief description of how a boat sails can help get non-sailors oriented. It can be as simple as a sailing circle, showing points of sail. Match the points of sail to times on the face of a clock, and post the diagram. When you are underway you can then explain, for example, that you are turning from 10 o'clock, past 12 o'clock, to 3 o'clock, when you are tacking from close hauled on starboard to a port beam reach.

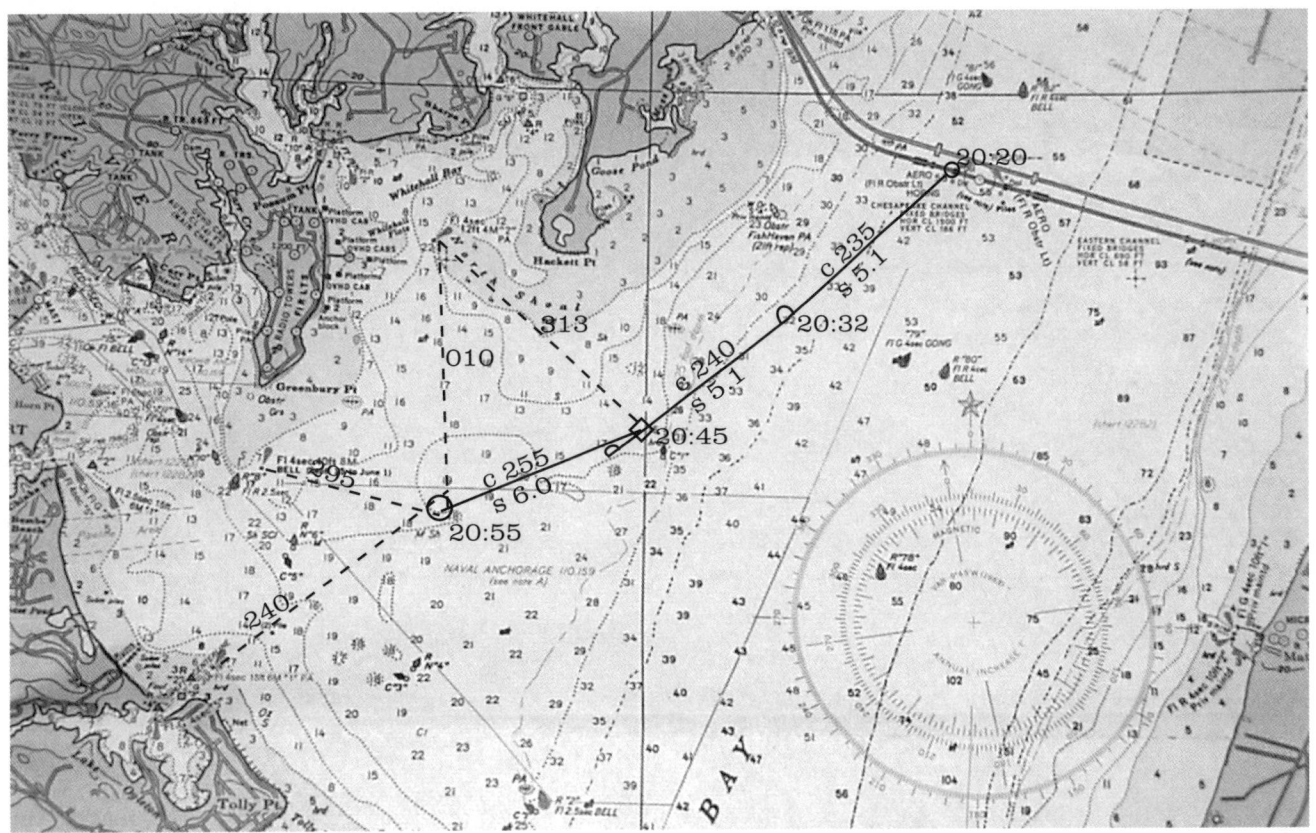

Here we see a DR position at 20:32, an Estimated Position based upon the DR and a single LOP at 20:45, and a Fix at 20:55.

2. Ded* Reckoning and Coast Piloting

You should navigate with a GPS. You do not need to know how to handle a sextant to consider yourself to be a sailor. Yes, GPS can fail, but you can also drop a sextant. You might consider carrying *two* GPS units - one fixed mount unit with large display, digital charts, remote antenna, and NMEA 0183 interface. The other a handheld unit. There are three reasons to have a handheld: for convenience, so you will have GPS even if the ship's power fails, and as an emergency unit if you must abandon ship. With the price of GPS in the low hundreds, and the cost of boats in the high thousands, redundant GPS is an option to consider.

Still, good navigators do not rely solely on GPS – or any other single type of navigation. They navigate redundantly, using GPS and checking their navigation on paper charts through coastal piloting (taking bearings on charted objects) and dead reckoning (tracking progress with compass, speed and time). It is really not so bad. In fact, piloting is an enjoyable part of shipboard routine.

Taking bearings, plotting positions, planning courses, correcting for current, anticipating and positioning to take advantage of changes in the wind. It is all part of the activity of cruising. It is how we stay busy while seemingly doing nothing all day.

---

* *Ded Reckoning*, as in Deduced, or *Dead Reckoning* as in accurate, like *dead ahead?*

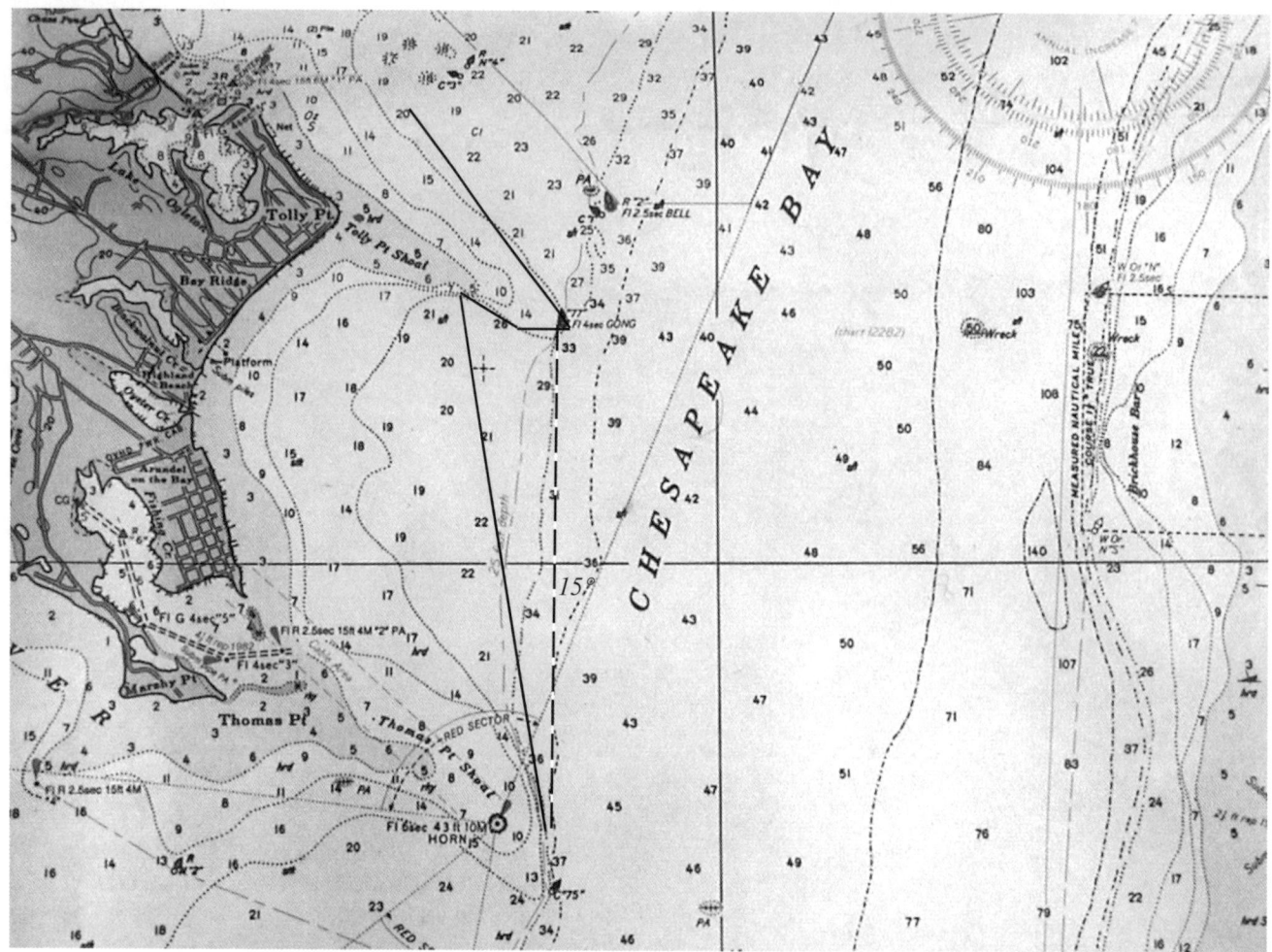

After passing Thomas Pt. Light in very low visibility, what course should we steer to clear Tolly Pt. safely? The rhumb line to the Tolly Pt. Gong is 15°. If we steer that course, then we might miss, and not

know it. A better course to steer would be to the left, inside the Gong, until we reach the 20 foot depth contour. We could then turn to starboard and follow that contour to the Gong.

Dead Reckoning and Coast Piloting

Deduced (Ded) Reckoning (DR) is navigation based solely on what the navigator sees on board the boat: the boat's course, her speed, and the length of time the boat has been under way. Together, they produce a direction and a distance run.

Coastal piloting is navigation within sight of land, aids to navigation (buoys, lighthouses, etc.), or other objects shown on a chart. The navigator plots the position by taking compass bearings on these objects and recording them on the chart as *lines of position* (LOP).

The best way to take a bearing is with a hand bearing compass. (Note: A great gift idea. The best ones are the size and shape of a hockey puck.) In lieu of a hand bearing compass you can sight across the ships compass, or aim your boat at the object.

Estimated Position and Fix

If there is a bearing on one object, then the spot where the LOP crosses the boat's track is called an *Estimated Position* (EP).

If there are bearings on two objects, where they cross is a *fix*. For accuracy, the bearings should be at least 60 degrees apart. A third bearing provides a check to assure accuracy. That's the most certain position.

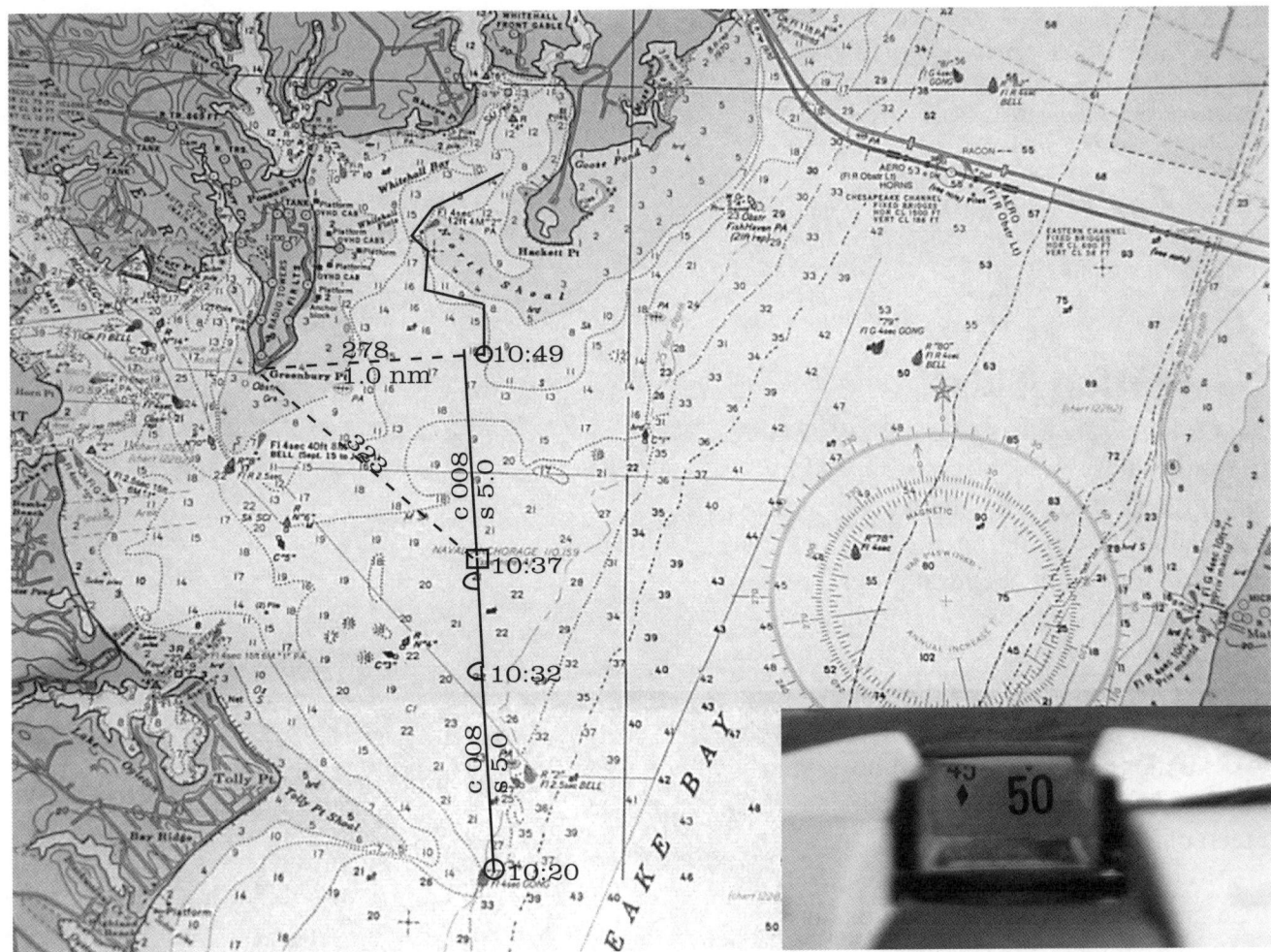

From Tolly Pt. we could proceed toward our destination in Whitehall Bay by Dead Reckoning. Catching a glimpse of Greenbury Pt. might allow us to get an LOP, and establish an Estimated Position at 10:37. Another bearing at 10:49 would give us a running fix, and with a doubling of the bearing, a distance off. If visibility ahead was still poor we

could follow the 15 foot contour to the North Shoal light. We might then choose to anchor behind Hackett Pt. until conditions improved. Of course, your fix is only as good as the bearings it is based upon. A hand bearing compass provides accurate and easy bearings.

Piloting Tricks - Running Fix:

When you can only get a bearing on a single object you get an EP. Advancing that bearing along based on your DR, and then combining it with an additional bearing, will give you a Running Fix. Here's the why and how: The two bearings give you two legs of a triangle. Your DR gives you a distance and angle for the third leg, which will only fit with the other two legs of the triangle at one position.

Distance Off:

Doubling the Bow Bearing is a technique to determine your distance from an object.

The distance you run in the time it takes to double the bearing equals your distance from the object. This method works for any doubled bearing, though narrow angles can be hard to measure accurately.

An example: If the object bears 45° off the bow, then your distance off will equal the distance run when the object is 90° off the bow (i.e. - on the beam). Of course, this info may be a little late if you are trying to make sure you stay a safe distance off. A 30°/ 60° measurement can give you a fix and allow you to plot a safe course earlier.

Aids to navigation – chart depictions:
Top: The Cow & Calf bell buoy. Red, #34, with a bell and a red flashing light with a 2.5 second interval.
Middle: Red bell buoy, #4, with a 4 second red flashing light. Also, unlighted can #3. It would be green.
Bottom: Falkner Island Light, flashing (white) every 10 seconds. 94 feet tall, and visible from 13 miles, conditions permitting.

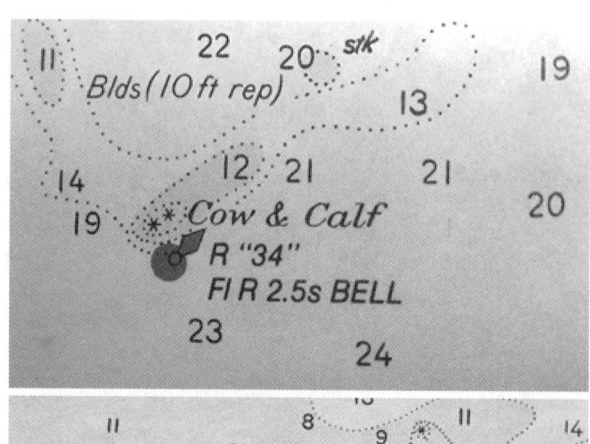

Navigation Aids*

Channels are areas of water deep enough to sail in safely, generally surrounded by unnavigable waters, and marked by aids to navigation, like buoys and lighthouses. Our lateral system of buoyage employs a redundant system of characteristics to indicate which side to pass on.

Red Right Returning is the fundamental rule. When going from a large body of water to a smaller one, keep the red buoys (and buoys with red lights) to starboard and the green ones to port.

Red buoys have different characteristics than green ones. Among unlighted buoys, nuns (which mark the starboard side of the channel when entering) are red, are pointed, and carry even numbers. Cans (which mark the port side) are green, are flat-topped, and carry odd numbers. The lowest numbers are at the outer end or beginning of the channel.

Lighted buoys follow the same pattern: Red lights and even numbers on the starboard side of the channel, green lights with odd numbers on the port side. Many buoys are also equipped with a bell (one tone) or gong (3 or more tones). The light patterns, called phase characteristics, vary so the buoys won't be confused with one another. Lighthouse phase characteristics also vary.

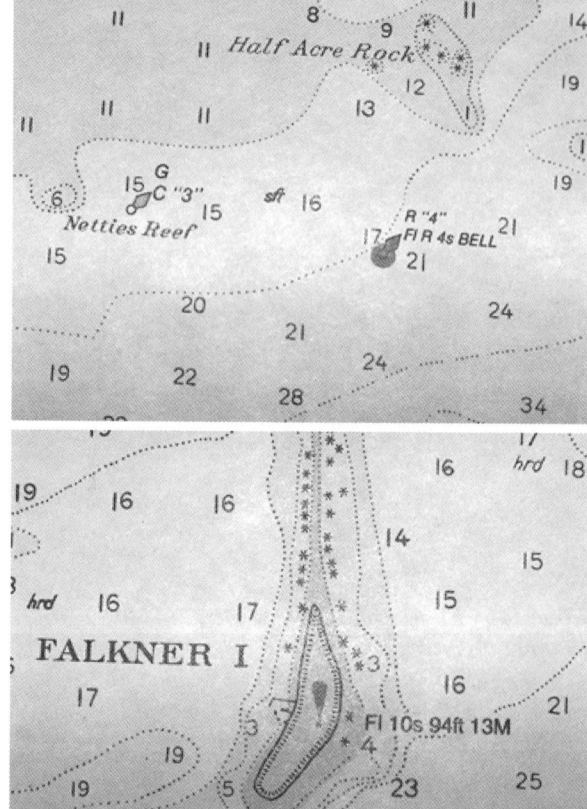

---

* The information presented here on buoys, channels, and aids to navigation is meant as an introduction. For a more thorough coverage of this topic consult <u>The Annapolis Book of Seamanship</u>, and other sources.

In addition to red and green buoys, there are other types buoys, each with its own characteristics. For example, there are buoys which mark the head of a channel, a split in a channel, or an isolated hazard.

The characteristics of each buoy and light are detailed on charts. NOAA Chart #1 shows the full list of symbols and abbreviations used on all charts.

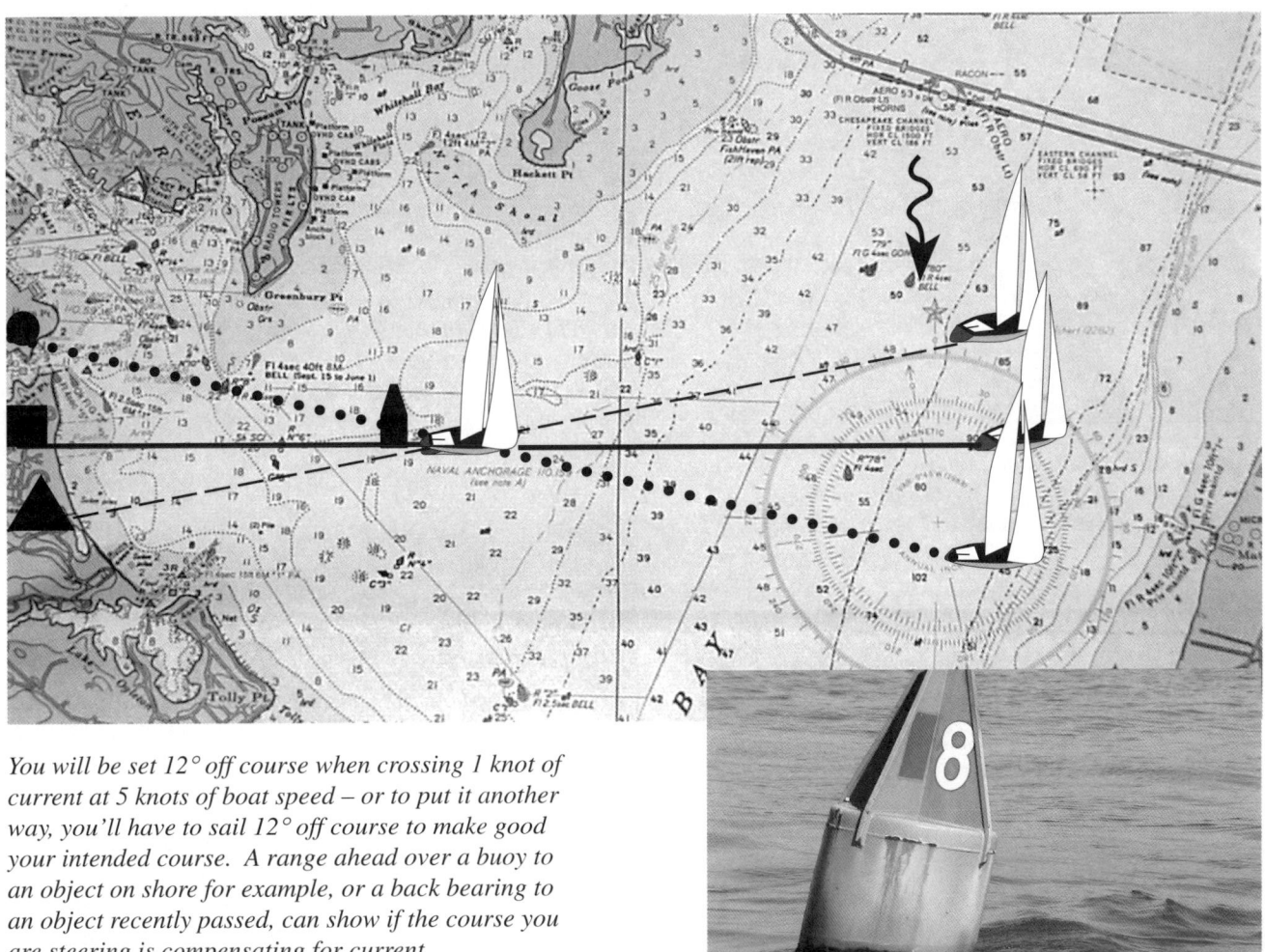

You will be set 12° off course when crossing 1 knot of current at 5 knots of boat speed – or to put it another way, you'll have to sail 12° off course to make good your intended course. A range ahead over a buoy to an object on shore for example, or a back bearing to an object recently passed, can show if the course you are steering is compensating for current.

Tides and Current

Using tide tables and charts, you can predict set (the current's direction) and drift (its speed). With this information you can estimate how currents will affect your boat over time. But beware the effects of wind on current, particularly in shallow bays. Strong winds can slow, accelerate, or even reverse tidal flow.

It's important to distinguish between two types of measurement of progress:
• Speed and course through the water. That's the readouts on the knotmeter and compass.
• Speed and course over the ground (SOG and COG). They include the effect of currents. You can calculate SOG and COG using coastal piloting and comparing the results with your dead reckoning.

Or you can read them on the GPS.

Current does not always flow in a straight line, especially near shore. It often twists around a point to form a back eddy. And of course tidal currents reverse as they cycle in and out.

To judge the effect of current, set up a range using near and distant objects, either directly ahead or astern, like a buoy and object on shore. If the two objects stay aligned, the range is "steady" and your course is compensating for current. A back bearing to a fixed object recently passed can also help you assess the impact of current.

If you are approaching an area of stronger current then set yourself up on the up current side before you get to that area.

Your pre-navigation should include a list of navigation aids you expect to see. If they do not appear "on schedule" check you position, course, and speed.

Use binoculars to scan the horizon for distant lights and objects.

Underway keep of log at regular intervals of position, speed and course sailed, weather observations, and anything else of note.

When you see a tug, look out for its tow. Do you know what they look like at night? At night the tug will show two (or three) white lights forward, one over the other. The stern of the tug would show yellow over white. If you see a tug's stern, beware – a barge is sure to follow!

Pre-Navigation

In preparation for cruising, particularly at night, you ought to undertake a whole collection of tasks related to navigation.

The first of these is studying the weather. Before setting out check that the weather is at least predicted to be suitable. There is reason to put yourself in harms way.

Next, you'll need to plan your trip, plotting and programming waypoints, and noting any shipping channels you will cross and other hazards along the way. Likewise, note suitable harbors and anchorages along the route which might provide refuge should you decide to cut the trip short.

Review the charts, intended course, hazards, and harbors with each the full crew. Review also the critical importance of keeping a log, and review the information and frequency with which it should be recorded. Secure charts and navigation tools for easy access.

In planning a longer passage, to the extent possible, avoid entering tricky or unfamiliar harbors at night. Plan your landfall for after dawn.

While underway tune in for weather updates on a regular basis, and record the forecast on a miniature recorder so you can review it.

With each watch change, the watch captains should review the log, weather forecast, sailing conditions, and expectations for the coming watch.

Log book entries should be made at regular intervals, or more frequently as necessary, including position, the speed and course sailed, weather observations, and anything else of note, including radio transmissions, and other observations. A regular set of procedures followed in easy conditions will create habits of particular value should conditions deteriorate or electronics fail - or both, as seems most common.

The more you can prepare, the more you'll be able to enjoy your passage.

Here we see three boats all with the same initial bearing to crossing traffic. The bearing widens for the black boat, and she can cross. Whew! The dark sailed boat sees a narrowing bearing – the ship will cross him. The constant bearing from the middle boat shows a collision course. Keep a lookout!

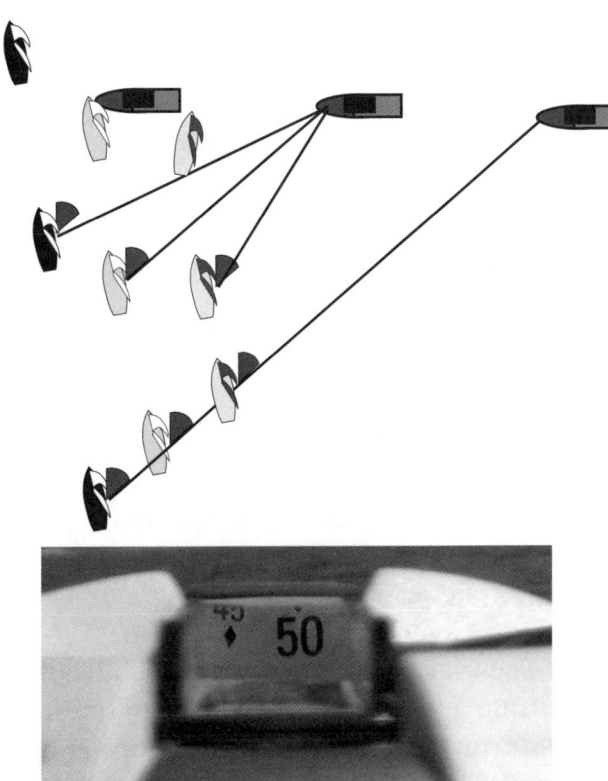

Seen from the bridge of a ship, a fleet of sailboats appears like so many gnats. If you are among the gnats, imagine the ship is full of deet, and steer clear.

Traffic

One thing you cannot entirely anticipate is traffic. You may expect there will be some, but when and where it will pop up you don't know.

Crossing with Traffic

You look up and see a tug and barge a couple miles away. At first it doesn't seem to be moving very fast because there's no wake. Minutes later you look again, and it seems suddenly much closer, and your courses are crossing.

What do you do?

• Take bearings on the traffic to estimate the crossing. A fixed bearing means a collision course. A widening bearing suggests you will cross; a narrowing bearing puts you behind. Don't try to cross if you cannot cross easily. Slow down and head toward the barge's stern.

• Radio the ship on channel 13, draw his attention to you, and let him know you intend to pass astern.

• This bears repeating: Slow down and head toward the barge's stern. Make sure you go behind the tug *and the tow!* You sure don't want to go between them.

Bow to Bow with a Ship

The next time you find yourself facing on coming traffic, consider the perspective from the bridge of the ship, and then decide what to do. From bow on, make a decisive move to one side or the other.

Make your intentions clear through an exaggerated change of course. You can put the ship's captain further at ease by contacting the ship on VHF channel 13, and telling the captain your plans.

If your courses are crossing, take the ship's stern. They pass quickly. Trying to cross can scare you to death even if you make it. On the other hand, I've yet to be hit be a ship I went behind.

Sailboat lights at night:

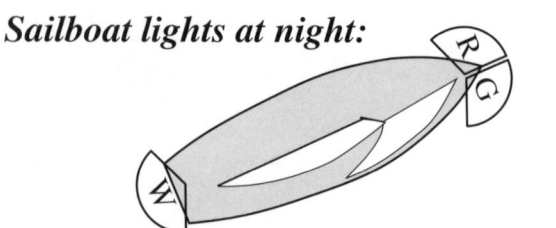

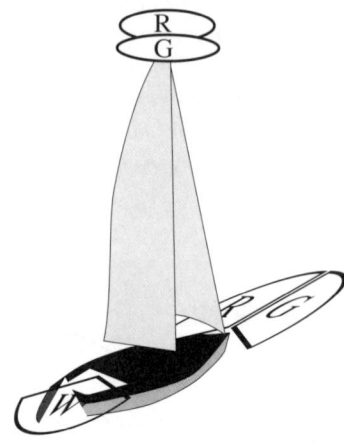

Red and green side lights and a white stern light or combination masthead tricolor

Sailboat with optional masthead all around red over green in addition to regular running lights

Traffic at Night*

Just as you do by day, you'll need to scan the horizon for traffic every 10 minutes at night. To see behind the jib or spinnaker head up and then bear off; or walk out to the bow – harnessed in – and scan forward and leeward from there. By day, we may see a boat. At night we see only lights.

Each type of vessel carries an identifying set of lights. Underway you will of course see only some of the lights on each boat.

Here are some things to watch out for:

Sailboats carry sidelights, red to port and green to starboard; and a white sternlight aft.

When approaching head on, alter course to starboard, and pass port to port. When converging, a boat on the right has right of way, so if you see the other boat's port red light, stay clear. If you are to leeward of a sailboat, and see forward lights, then assume they do not see you behind their headsail. For that matter, never assume anyone sees you.

Of course, there aren't just sailboats out at night. On the following page are diagrams of *some* of the lights you might see.

Any vertical arrangement of lights deserves special attention. White over white is the forward side of a tug – you should see sidelights as well. If you see a yellow light over a white light, you are seeing the stern of a tug boat. Lookout – the barge, which carries sidelights and a stern light, like a sailboat, is sure to follow!

Also, give a wide berth to commercial fishing boats – showing red over white, and trawlers – showing green over white.

Hailing and Communications

Once you see traffic, what should you do? Steer to keep clear, for one. Make your intensions clear through an exaggerated change of course. You can also communicate your intentions over VHF radio.

The trick in using the VHF is to identify yourself, and the vessel you hope to speak with. You hail ship to ship at low power on Channel 13. Channel 67 is the alternate. Some still use 16 for hailing, though it is prescribed for emergency use only

Describe yourself, your position, and the position and course of the boat you are hailing. For example: "This is the sailing vessel, *Lumpy Gravy*, eastbound on Long Island Sound south of New Haven harbor hailing the tug with tow headed north approximately 5 miles from New Haven."

Once you establish communications, then state your intentions: "We see your course, and intend to alter course to pass astern of you and your tow."

---

* The information presented here on navigation lights and rules of the road is meant as an introduction. For a more thorough coverage of this topic consult <u>The Annapolis Book of Seamanship</u>, and other sources.

A few examples of ship's lights at night:

A power boat up to 50 meters adds a white light forward to the sailboat lights. This applies to sailboats under power as well. Under power the masthead tricolor is not permitted.

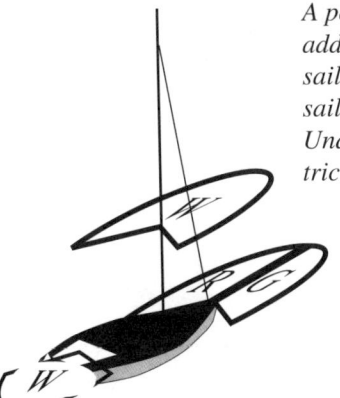

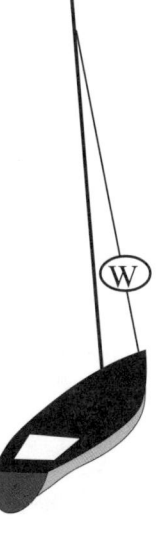

At anchor show an all around white light from the masthead or where it can best be seen.

Ship 50 meters or more, (often used on ships less than 50 meters) show red and green sidelights, sternlight, and two additional white lights – one forward and the second higher and further aft, forming a range. Look out!

A tug towing a barge carries red and green sidelights, two additional white lights forward, one over the other, and a yellow light over the white sternlight. Add a third light forward if over 200 meters. The barge would carry sidelights and a sternlight, like a sailboat.

At night you would see some but not all of the lights depicted here.

---

* The information presented here on navigation lights and rules of the road is meant as an introduction. For a more thorough coverage of this topic consult <u>The Annapolis Book of Seamanship</u>, the Coast Guard's booklet: <u>Navigation Rules: International-Inland</u>, the <u>COLREGS</u> (International Regulations for Preventing

---

To sail to a destination directly upwind you must sail 1.4 times the straight line distance. For example, to sail to a destination 10 miles upwind you would have to sail 14 miles; 7 miles on port tack, and 7 miles on starboard. You could sail all port first, or all starboard first, or you could tack back and forth. In the end you'd sail 7 miles on each tack.

When boats are equally far upwind then they would lie on the same line perpendicular to the wind - an LEP. Though spread out laterally, boats on the same LEP would have the same sailing distance remaining to reach an upwind destination. Their remaining time on each tack might differ, but their total remaining sailing distances to the destination would be the same.

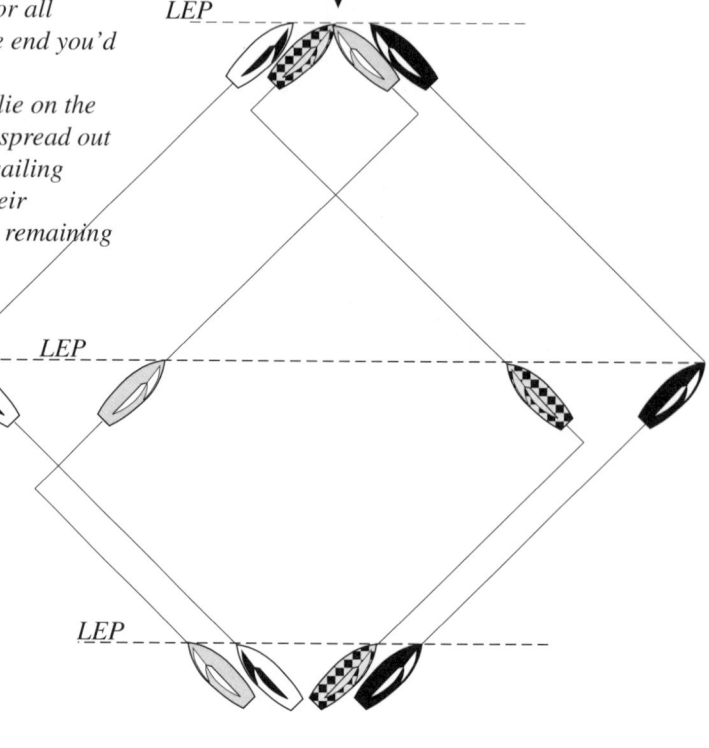

3. Sailing in Wind Shifts

Our efforts to navigate a safe passage are compounded by the fact that (too) often we cannot sail directly to our destination. If the destination is upwind, we'll have to tack to get there. Our efforts are further complicated by the fact that the wind is rarely steady. That's the bad news.

Now the good news: By taking advantage of windshifts, you can reduce the distance you must sail to reach an upwind destination. The wind is rarely steady. Usually it is shifting – either shifting back and forth – *Oscillating*, or shifting gradually in one direction – *Persistent*. Persistent shifts are either veering (clockwise), or backing (counter clockwise).

As the wind shifts, it changes the course you can sail. A *header* is a shift which heads you away from your destination. A *lift* is a shift which lifts you up toward your destination. When one tack is headed, the other is lifted.

No Shifts

To understand the impact of wind shifts, and how to take advantage of them, consider first boats sailing upwind without any wind shifts. In the figure above four boats set out toward an upwind destination. Assuming they sail at the same speed and angle, they

When the wind shifts, one tack is Lifted (up towards the destination) and the other tack is Headed (away from...).

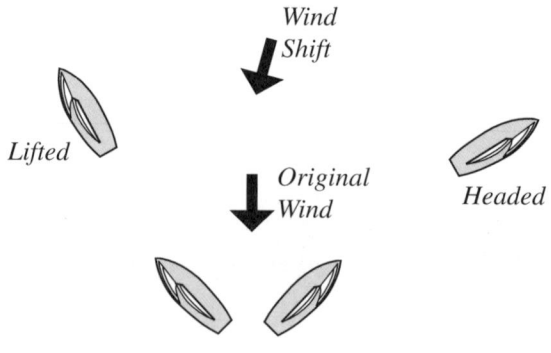

remain equally far upwind, though they are spread out side to side. They are spread out on a line perpendicular to the wind - a *line of equal position* or *LEP*. Although the straight line distance to the destination may be different, the sailing distance remains the same for all boats on the same LEP.

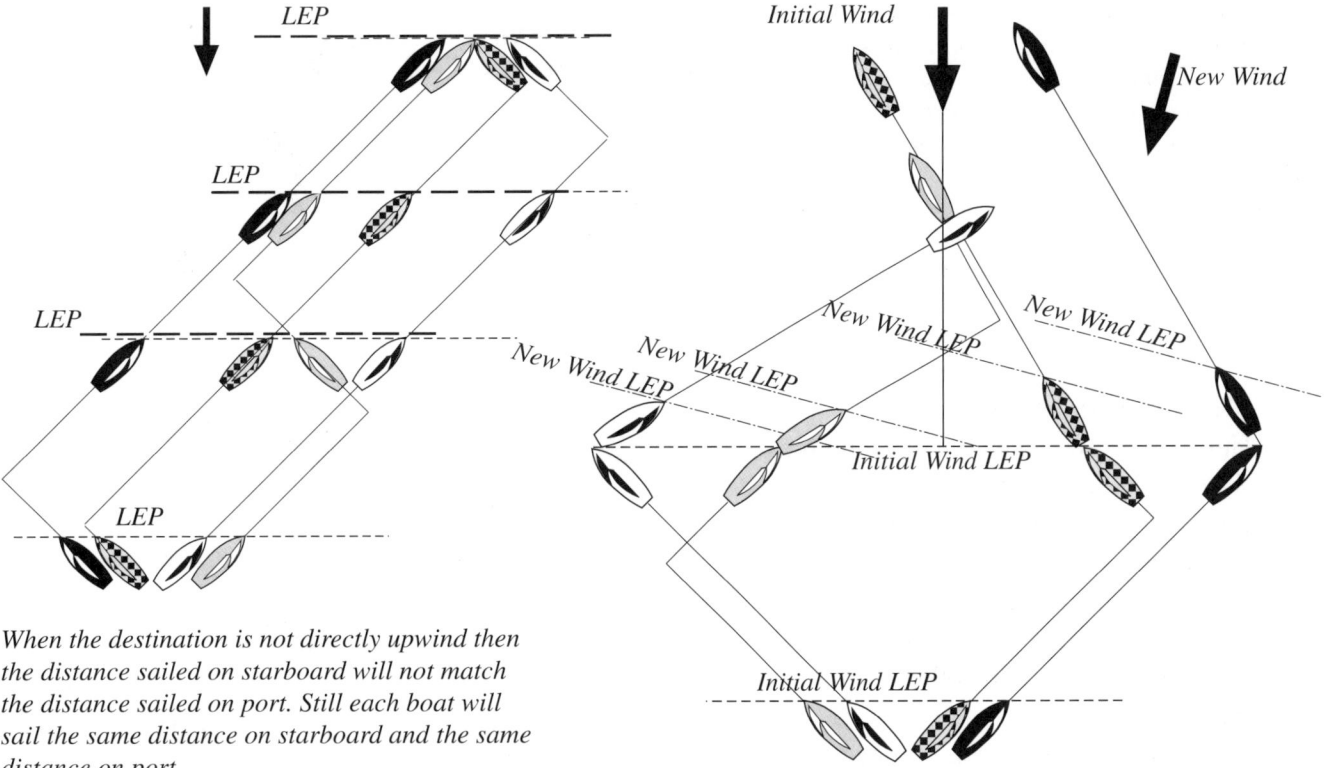

When the destination is not directly upwind then the distance sailed on starboard will not match the distance sailed on port. Still each boat will sail the same distance on starboard and the same distance on port.

When the wind shifts then boats closer to the shift are at an advantage, while boats away from the shift have their sailing distance increase. In this example a shift to the right helps the boats to the right. It may not look like much here, but the advantage in even a small shift can be dramatic – measured in miles, not feet.

To reach an upwind destination in the absence of wind shifts they will have to sail a fixed distance on starboard tack and a fixed distance on port tack. Each crew has a choice - sailing port tack first, and tacking once to starboard - like the black boat, or sailing starboard first, and then sailing port - like the white boat, or tacking several times, like the grey and checkered boats. In the end, without wind shifts, they each sail the same total distance on port tack, and the same total distance on starboard tack.

If the destination is directly upwind then the distance on port will match the distance on starboard. If the destination is not directly upwind, but you must still tack to get there, then the distance on port will not match the distance on starboard. Still the sailing distance on starboard remains fixed, as does the distance sailed on port, even though they are not equal.

Likewise, as long as the boats lie upon the same LEP the remaining sailing distance to the destination is equal.

The Wind Shifts!

When the wind shifts then the remaining sailing distance changes, and the LEPs rotate – the LEPs are (always) perpendicular to the new wind. Boats closer to the new wind direction have their remaining sailing distance reduced, while those on the side opposite the new wind now have further to sail!

If the wind is going to shift, then sail toward the new wind. In the example above the four boats set out toward an upwind destination and when they are about half way there the wind shifts to the right. The boats on the right benefit, while boats to the left are set back.

By taking advantage of windshifts we can shorten our sailing time or expand our sailing distance upwind. To take advantage of windshifts sail toward the new wind. If the wind is expected to come from the left, go left. If the wind is expected to shift to the right, then go right.

If the wind is oscillating back and forth, as shown here, then tack back and forth and sail on the lifts. Each tack puts you on the lifted tack, and takes you toward the next shift.

More on Wind Shifts

Here are a few ideas to help you take advantage of wind shifts.

1. Sail the course that takes you closest to your destination. Note the direct compass course to your destination, then sail on the tack that aims you closest to it. If the wind shifts to make the other tack closer, then tack.

2. In oscillating winds, tack on headers. If you're headed more than about 10 degrees for a few minutes – meaning that a wind shift makes you steer a course that's lower, then tack and sail lifted. When that tack gets headed, tack again. Sail lifted.

3. Sail towards the new wind. In a persistent shift, with the wind shifting gradually in one direction, or when you expect a shift (based on the forecast, say), then sail toward the wind you expect.

Of course, part of the trick is predicting what the wind will do. There are a number of clues which can help you predict wind shifts:

1. Trends you observe. As you sail, you can keep track of your close hauled compass course, and look for trends. You can also see the wind on the water.

2. The weather forecast. The forecast can tell you to expect gradual shifts. As weather systems and fronts move, the wind shifts in a predictable way.

3. Thermal "sea breezes." In many areas when conditions are right, you can anticipate the arrival of a thermal sea breeze blowing onshore as the day heats up.

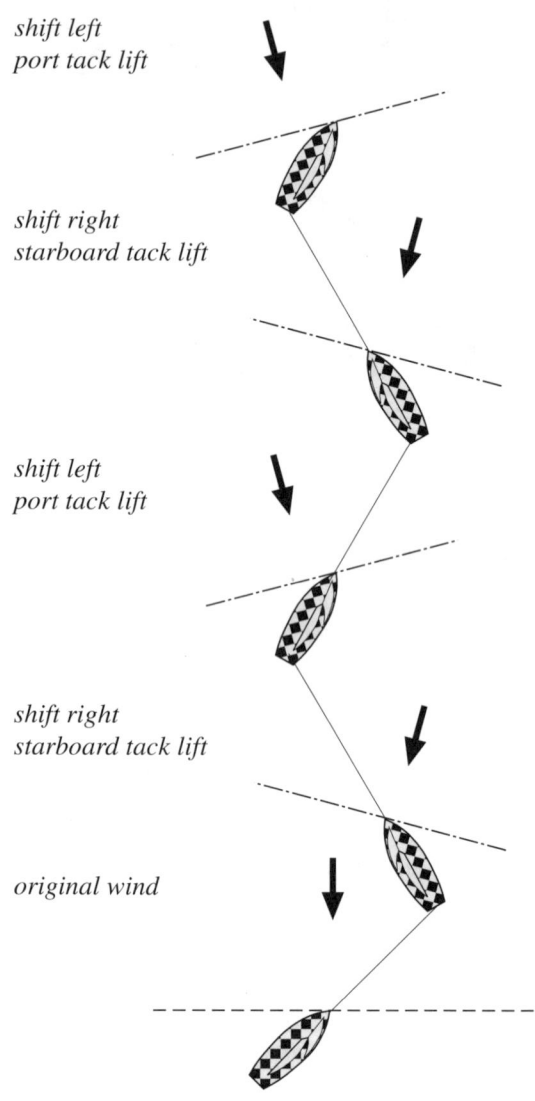

shift left
port tack lift

shift right
starboard tack lift

shift left
port tack lift

shift right
starboard tack lift

original wind

These same principles apply downwind, only in reverse. When sailing on a run, jibe on lifts – the opposite of tacking on the headers.

When all else fails, if you don't know what the wind will do next, (imagine that!) then sail the course which takes you closest to the destination.

Of course, if your destination is upwind, you could pick a new destination, or you could just motor straight to it... pounding and splashing, wasting precious fossil fuels, unable to converse above the din of the engine. no No NO – we are sailors! We only motor when it is calm.

No Sunscreen Needed:
Sailing at night let's you escape the sun and expand your cruising horizons.

4. Sailing at Night

Extraordinary Beauty, Extra Caution

Sailing at night is one of the great pleasures of cruising – a unique experience offering great rewards but requiring an extra measure of preparation and caution.

Just what are the rewards? Night sailing puts you in touch with your senses and your world in ways that day sailing does not. No longer able to rely so heavily on our sight, we turn to our other senses. We feel an approaching front in the changes in humidity, temperature, strength, and direction of the breeze. We feel the trim of the boat through its motion and power.

The lights from shore and ships, the sounds of the sea and of our boat. Our attention turns to these as the regular sights of day sailing fade to black.

Yet in the darkness we marvel at how much we can see. As our eyes adjust we can see all we need to see on deck, and we marvel at the density of stars in the night sky. Or perhaps we can follow the moon across the sky or sail beneath a blanket of darkness as black as black can be.

And then of course there is the sunrise. The first loom of light – earlier than we might expect, and the glow spreading across the eastern sky. Eventually there is a bright orange line across the eastern horizon, and then the sunrise itself. Or perhaps a red sky, as the sun shines up into approaching clouds, and all that that portends. ("Glad we made our passage over night. Our landfall will take us to a snug anchorage where we can catch up on our sleep and ride out the weather.")

All that, plus night sailing expands our cruising grounds over new horizons while affording an escape from the glaring sun.

And what of the cautions?

Night sailing requires preparation of the boat and crew, as well as advance planning, to assure a safe and enjoyable passage.

Equipment

Aside from the regular compliment of daytime safety gear, some additional items are called for:

Lee cloths or adjustable bunks which allow crew to sleep securely when heeled and underway.

A gimballed stove.

A spotlight. Either self contained, battery powered, or a 12 volt light with a long cord. A spotlight is great for finding unlighted buoys, as well as other hazards. A spot light is also critical in an emer-

Squall drill: Practice calling all hands to stations to quickly shorten sail. Run through the drill by day, and again at night. Spare yourself a mutiny by calling the drill at a watch change.

The squall line pictured here would be hard to miss during the day, but it might be on top of you before you see it at night. Changes in the night sky, and sudden changes in wind and temperature usually provide some warning.

gency to show off your boat to traffic, and essential for night time crew over board rescue.

Cyclic glow sticks. Bend 'em and they light up. Great to put a glow on a compass if the compass light fails, and for other similar purposes. Keep one on a lanyard in your pocket, and twirl it overhead if you fall overboard.

Red cabin lights. Some boats have duel action lights – throw the switch one way for white, the other for red. Mark the switches so you can tell. Red lights are critical at the navigation station.

It is remarkable how much you can see once your eyes adjust to the darkness. Keep lights to a minimum to protect your night vision. For example, close the companionway before lighting cabin lights. And use you binoculars to scan the night horizon. You'll be able to see lights and objects you would otherwise miss.

For daytime sailing the main halyard should be marked for each reef position. For night sailing that same mark should be a whipping so you can feel it as well as see it.

A masthead tricolor light is dramatically more visible at a distance than deck level running lights. Deck level lights are more visible to boats nearby. And you're not allowed to use both. If you do have deck level lights, then you are allowed to add a masthead red over green to identify you as a vessel under sail.

Boat Preparation

As the sun sets take a twilight tour. Make sure halyards are clear to run. Check the jack lines, stow any loose gear (above and below deck), fill a thermos with hot water or soup, issue flashlights, and charge the ship's batteries.

The boat should be set up for a quick reef, with reef lines and halyards in place. Each person should have a preassigned position and task if "all hands" are called to shorten sail at night. An afternoon squall drill will help prepare the crew. Rerun the drill in the dark to be sure you're ready to shorten sail on a moments notice in the dead of night. You can spare yourself a mutiny by calling the nighttime drill at a watch change.

Personal Preparation

Dress well, and stay warm. Organize your personal gear so you can find what you need in the dark without disturbing the off watch.

Don you harness if you don't have it on already, and clip on. Make it a rule, and lead by example: Clip on when sailing at night.

An afternoon nap is a good idea if you plan to sail through the night. Which brings us to:

Night cruising: Clip in when on deck. ^

Rest when you can. In anticipation of an overnight sail, try to get some rest during the day. Note the lee cloth rigged to this bunk which would allow comfortable sleep even when the boat is heeled. >

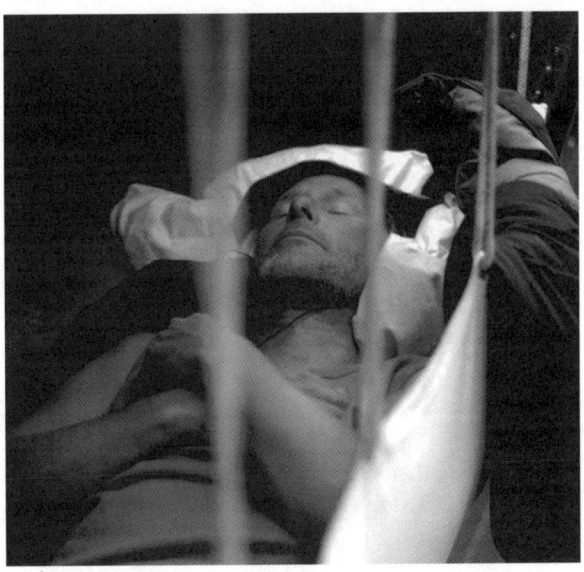

Watch Systems

There are a number of options when it comes to setting a watch. Select the system best suited to the number, skill, and temperament of your crew. Choices include:

• The traditional watch system has two watches at four hour intervals. Some crews prefer to stand three hours during the overnight. On an extended cruise 4 hours by day, and 3 by night, is self dogging – that is, it rotates the watches each day. If you stick with a four hour watch, then the 1600 to 2000 watch can be divided in two – 1600 to 1800, and 1800 to 2000 – to dog the watches.

• There are many other options. Some crews prefer a rotating watch system. Rather than divide the crew into two groups, the crew is divided into four, with one team going off and another coming on every two hours. This brings fresh hands, and provides more continuity than turning the entire crew every four hours, but can be disruptive to those below.

• Some crews divide into pairs, with each pair having comparable skills. Every hour one person comes on and their partner goes off.

• Another option is to divide the crew into three groups. With a crew of six, for example, one pair would be *on watch*, the second pair would be *off*, and the third pair would be *on call*. In easy conditions the *on* watch handle the boat themselves, allowing the *on call* watch to get extra rest. In more challenging times the *on call* team is called, putting two thirds of the crew on deck. This system affords extra rest when possible, and provides more hands when needed.

Regardless of the system you use, it is critical for crew to get rest. Many crews, and particularly those new to cruising, have a hard time settling into shipboard routines and getting rest early in a cruise. As skipper, encourage crew to rest when they can, and to nap during the day.

It is also important to establish *when to call* guidelines to wake the skipper. Assure your crew that you will rest better if you are confident they will not hesitate to call if:

• the wind shifts, or the sky or temperature change dramatically.

• your position is uncertain, either due to a discrepancy in your piloting versus your GPS position, or due to the appearance of an unexpected light or landmark.

• anything out of the ordinary that makes the crew uneasy.

Pulpit mounted running lights are far less visible than a masthead tricolor. When the boat heels assume that only fish will see your leeward sidelight.

Even with the help of a flashlight sail shapes can be difficult to see at night. This photo shows a genoa with a reflective draft stripe and the moon just beyond the forestay.

Sailing

The *sailing* part of sailing at night offers its own challenges. Steering at night can be tricky in the absence of a horizon. Look up when steering – avoid steering strictly by the compass. It is difficult, and no fun. Steer instead to a star, a navigation light, or to the trim of sails and feel of the boat, and reference the compass periodically to see that you are holding course.

Sail trim presents its own challenges, as sail shape is difficult to see, even with a bright flashlight. Set your sails with extra twist to offer a wider steering groove, and test trim periodically. Ease the sails to a luff, and trim in.

In the absence of many of the usual visual clues to performance, watch for changes in apparent wind angle, wind speed and in the feel of the boat to prompt you to check trim.

In general it is best to adjust jib trim first, and then adjust the main to match. Try to keep the middle telltales on the jib flowing, and do the same for the leech telltales on the main.

When sailing with an auto pilot it is often easiest to hand steer to get the boat on course, and then reengage the pilot.

Cruise at night. Prepare yourself, and go.

Conclusion

Before you head out for a full overnight take a midnight sail in your home waters. Test your gear, and yourself. Practice sailing, navigating, and reefing in the dark.

Preparing for night sailing may seem like a lot of trouble, but it is worth it. It is worth it as an end in itself – just to be out in the night, and it is worth it because it opens up your cruising territory. Sailing for a day, a night, and a day will quickly take you far from home, and leave you more time to explore new cruising grounds.

Go. When the weather looks right, with enough good crew, go. Don't forget to count the stars, look out for satellites passing overhead, watch the phosphorescence glow in your wake, listen to the sea, and breath the night air. Tomorrow you can sleep.

5. Problems and More Problems

As we've said, very few problems pose an immediate threat to life, limb, or boat. For those few that do we must prepare. At the same time, you can get into serious trouble through the accumulation of small things. It is no one thing that gets you, but many smaller things piling up. It is best to keep after the small things. A regular, routine inspection of the boat and its systems can turn up minor problems with easy solutions before they become something more. While we've explored the big problems earlier, we'll take some time here to look at smaller problems, and see how to deal with them before they accumulate.

We'll start with a page of advice on engines, and then look at a selection of electrical and mechanical woes, as well as a series of other problems which can plague you on board, or under way. Finally, we'll try to tackle some of the most gnarly problems – those you may confront ashore.

As troubles arise during a cruise here are items to address:

- Is there any immediate danger?
- What is the best immediate response?
- What is the best solution?
- How can we prevent recurrence?

Electrical and Mechanical

Electrical system down

You've been running the engine regularly to keep the batteries charged – but the lights seem dim.

- Check the amp hours and charge.
- Check the battery switch.
- Turn off all low priority items.
- At night, rig an alternate anchor light.
- Check for loose wires in the charging loop and at the battery terminals.
- See if lights brighten with the engine running. If not, then the system is not charging.
- The alternator belt may be loose.

Your boat should have two isolated battery banks – one for "house" use, and the other reserved for engine starting only.

Engine won't crank over

If the engine won't crank over then:
- The batteries are too weak to turn it.
- The starter has failed.
- The connections are loose or corroded.
- Set your battery switch to *Both*.

You may be able to get the engine to turn by releasing the compression valves on the engine. If the battery and starter can't manage that, then you may be able to hand crank a small diesel engine to get it going.

Study your engine at leisure to know how the compression fittings release.

In the end you may end up cruising without power – electrical or mechanical. It is not as bad as it sounds.

Engine cranks, but won't start

If the engine cranks, but won't start, then you can surmise a fuel problem. It may be:
- Air in the fuel line.
- Water in the fuel.
- Clogged fuel line or filter.
- Out of fuel! (Oops)
- The kill switch is not fully released
- Or ...

Bleeding the fuel line of air is one of the joys of cruising a sailing auxiliary. Welcome to the fraternity! Pull out the owner's manual, study the process, find – and mark with yellow paint – the bleed valves and hand pump for clearing the fuel line, in anticipation of this day.

When the fuel is low, or after a rough passage, debris in the bottom of the tank can clog fuel filters. Carry spare filters and change them. Also, install a fuel separator to clear water. Fuel separators which clear water during the fueling process can prevent troubles down the line.

No Water Coming Out of the Exhaust

Shut down! If you are in a crowded spot you can motor briefly out of danger, and anchor or set sails.

• Check the water intake - it may be clogged.

• If the boat is heeled while motor sailing the cooling water intake may lift out of the water, and the engine can over heat. Make sure water is pumping out the engine exhaust when you are heeled.

• The water pump impeller may have been damaged as well from running dry.

• The impeller may be damaged from picking up debris.

Checking and changing the water pump impeller is a good place to start. Carrying spares (plural!) is more than a good idea.

Engine Quits

If the engine quits underway, the first guess is a fuel problem. Especially when motor sailing, pitching, or heeling, air or debris can get in the line.

To the *cranks but won't start* diagnostics on the previous page add:

• A line wrapped on the prop (more later)

The Engine Does Not Engage in Forward

Linkage! Try reverse. Sometimes you can at least go backwards. Have your sails ready to go. It is prudent to keep the jib ready to unroll and main halyard on whenever you are under power.

An engine, at left, showing three compression levers. A raw water strainer, center, and spare impeller, right. Here's a list of spares to carry:
• *Impeller*
• *Oil filter (and oil)*
• *Water filter*
• *Fuel filter*
• *Hose clamps (assorted)*
• *Hose (sized to your engine)*
• *Alternator belt*
Here's a thought: To keep your impeller and belts from taking a shape during the winter pull them off. In the spring install a new impeller and belt, and save the old ones as spares.

First get out of harms way. When you get the chance, test the gear shift with the engine off. Open the engine box, and see if the gear shift cable is working properly. Most of these problems involve the cable from the shift lever to the engine. You might also find that the neutral button is stuck.

The trouble may be with the transmission itself. Shifting at high RPM can lead to damage. When shifting, throttle back... wait for the engine to idle down, ... and then shift. Easy to say here, tougher to do in a tight spot.

Engine troubles are a nuisance, but rare is the cruising sailor ready to give up the convenience of motoring through a calm, or charging batteries for the ever growing number of electric gadgets. Do regular maintenance, read your owners manual, and don't forget to change the oil.

Don't rely on shackles when you go aloft. Use knots, and two halyards (if you have 'em available).

Be careful not to get conked in the head when diving to clear a line off the prop. A prop in a aperture, such as this, is less likely to foul than an open set up between a fin keel and spade rudder.

Problems On Board

You Loose a Halyard Up the Rig.

Eenie, meenie, miney, mo,
you're the lightest, up you go!

Going aloft under way, even in small seas, can be difficult and unpleasant. Wait until you can get to a protected location. You need a secure bosun seat and halyard, even if you have mast steps of some kind. Ideally when you go aloft you should go on two halyards as protection in case of trouble with one. Do not rely on halyard shackles. Use a bowline and bypass the shackle.

Climb as best you can, with the crew taking up as you go. Don't expect the crew to drag you up the rig. You've got to assist by climbing.

If you can recruit many hands, then the lift can be made much easier. Two people working each halyard, for a total of four pulling, makes light work.

While you are aloft, inspect the rig and lubricate halyard sheaves.

A Line Gets Fouled in Your Prop.

Oops.

Eenie, meenie, miney, mo,
don the mask and down you go!

Don't try to restart the engine; you could end up tightening the wrap or damaging the shaft or shaft strut.

Do carry a dive mask on board for this and other reasons, including checking the anchor set, inspecting and cleaning the hull, and for fun.

If the line is from on board, then you may be able to work it off by pulling firmly. Be careful as you work the line around. For example, don't try to pull it around the stern to pull from the other side, as you may foul the rudder in the process.

Diving to the prop is not too hard in a calm anchorage, but it can be quite hazardous in waves, as the pitching hull can knock you out cold. Sail to a calm place and anchor before you dive. And be sure to shut down the engine.

When you dive, try to work the line free without cutting it. If that won't work, then dive with a knife on a lanyard, and be careful.

^ Tie jib sheets with short bowlines which won't snag, and with long tails which won't shake out.
> This reefed main may tear at the slug as the reef point and tack hook set back are not well matched.

The Knots on the Jib Sheets Come Undone

Turn downwind and roll the sail up (as long as you can reach the rolled clew), or lower the sail. To lower the sail, first turn to a reach, so the sail luffs to leeward while you go forward. When lowering, turn upwind so the sail luffs over the deck. Sit on the deck at the tack, and pull the sail down.

Tie better knots, and rehoist.

Missing Batten

Battens in the mainsail and small jibs may fly out if the sail luffs. Inspect them regularly and sew them into the sail. If you do lose one, replace it promptly. Carry a spare or two on board - or a length of batten stock which can be cut to length. (Be sure to file the ends smooth or tape them over after cutting.)

While a missing batten may seem like little more than a nuisance, sailing for an extended time without a full set of battens will shorten the life of the sail.

A Torn Sail

If your sail tears, take it down and repair it promptly before the tear can grow. Carry a sail repair kit for this purpose, consisting of:
- sticky back dacron
  - one 6 inch by 6 feet roll
  - two 2 ft x 3 ft sheets
- 1 inch sticky back kevlar
- needle(s) - several, some curved
- sewing palms
- waxed thread
- scissors
- seam ripper

Most genoa damage is caused by the sail backing against the spreaders or chafing on the stanchions. Proper tacking technique, along with spreader and stanchion patches, can prevent this damage.

Mainsail damage often occurs around batten pockets. Excessive luffing can accelerate wear and lead to damage. Full battens reduce luffing, but require periodic inspection at the inboard end, where the batten meets the luff of the sail. The other common source of mainsail damage are overloading slugs or slides, and chafe on the spreaders, either at full hoist or when the main is reefed.

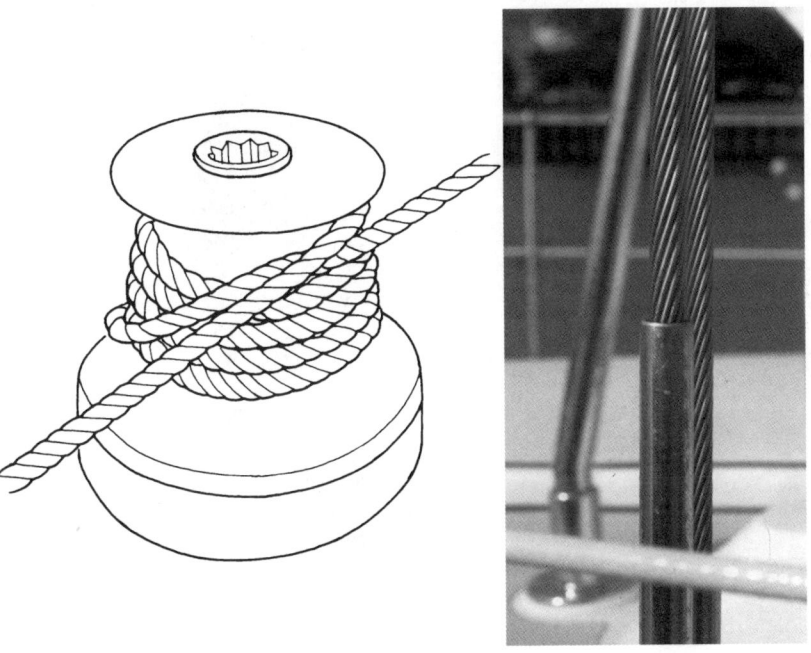

Winch Overrides

If there are more than two turns of a sheet, halyard, or other line on a winch before the load comes on, one turn may very likely ride over another and jam the line. To remove the override, take the load off the line forward of or above the winch and straighten out the turns. If it's a jib sheet, rig and tension another sheet to a winch. If it's a halyard, take the load above the winch. Or using a spare line (called a "short sheet"), tie a rolling hitch to the fouled line forward of or above the winch, then pull on the short sheet.

You can often take the load off a jib sheet by heading up and luffing the sail, or by lowering the jib halyard a few feet.

Standing Rigging Parts

If you lose a windward shroud, luff sails immediately, and then tack to take load off. Similarly, turn downwind if you lose the forestay, and upwind if you lose the backstay. Run spare halyards and the spinnaker pole lift as temporary supports.

Bulldog cable clamps and a short length of rigging cable can provide a temporary fix.

Roller Furler Problems

Roller furler problems are much less prevalent than they used to be. Better engineering, and better understanding of the nuances of set up of the drum and halyard have solved many problems.

If a roller furler jams, first reverse what you've been doing. If that doesn't work, check the control line for kinks in blocks and overrides on the drum.

Tightening or easing the halyard or headstay can sometimes alleviate problems as well.

If the roller is still jammed, look aloft to see if the halyard has fouled in the sail or on the top swivel. If nothing works, wrap the sail around the headstay. You may have to sail in circles initially to roll the sail.

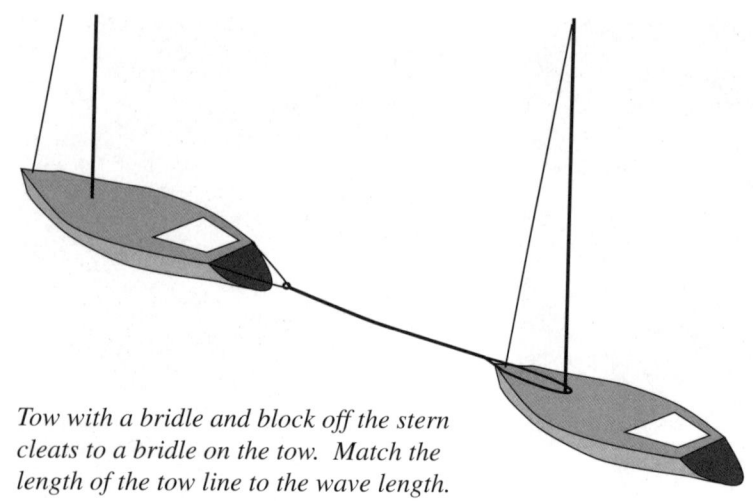

Tow with a bridle and block off the stern cleats to a bridle on the tow. Match the length of the tow line to the wave length.

The only trouble with fishing is that sometimes you catch a fish!

Towing

A disabled boat needs a tow. How do you set it up?

The tow line should be tied to a strong part of the boat, like the windlass, a reinforced bow cleat, or a winch, or around the mast. Tie to a couple of spots with a bridle to spread the load.

The tow boat should use both stern cleats and a bridle with a rolling block to allow steering and to balance the load. Do what you can to eliminate chafe. Set up a communication system with the tow boat. Use a horn to get attention over the noise of the engine, and hand signals for faster, slower, and stop.

The tow line should be long, nylon, and matched to the wave length so the boats are in matched positions on successive waves.

Land a Big One

You are trolling while surfing downwind with a cruising spinnaker on a glorious day and you hook a really big fish. Now what?

Douse the spinnaker with the snuffer, and heave to while you try to land your catch. Stow the fish in a bucket or cooler until you reach your anchorage. Then put a crew member in the dinghy to clean the fish. Pass the filets aboard, and clean the dinghy.

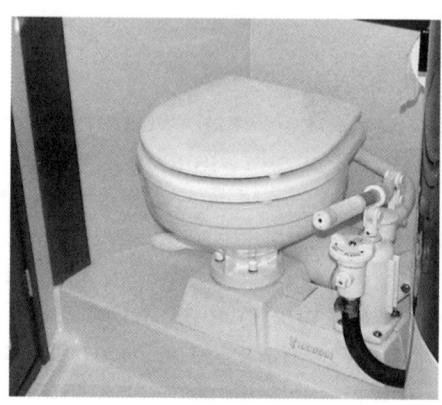

*There's lots of things, like pins and strings,
filter tips, and other things,
which shipboard plumbing will reject,
so my dear, be circumspect.
It's all a bother, and a care,
but oh my dear, so necessaire'*
only tissue!

The Head is Clogged!

Ah, the joys of yachting!

Maybe the holding tank is full.

An ounce of prevention is critical here. Take all new crew through a thorough lesson on head operation. In addition to posting detailed instructions on head use, you might add the poetic offering, quoted above, or the more direct:

Nothing in the head (save toilet paper) if you didn't eat it first.

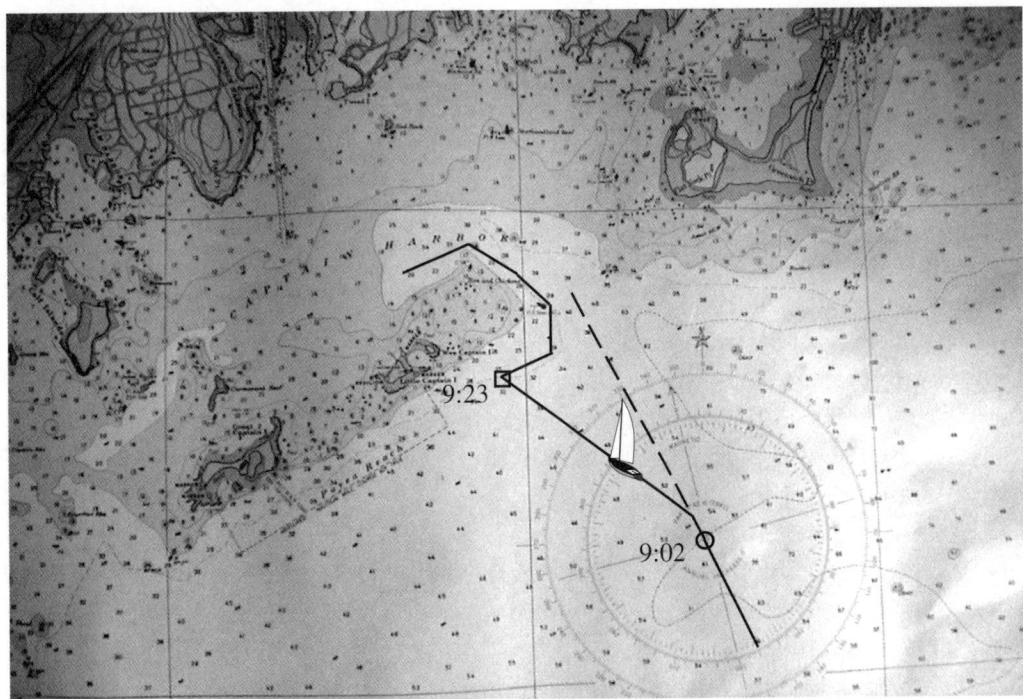

A Wind Shift

You're sailing along pleasantly on a nice reach in smooth water and favorable current when Aeolus, king of the winds, decides he doesn't like you and unleashes a wind right on your bow. Suddenly you're banging directly into a rough sea on a dead beat.

What can you do? Consider these possibilities:
• Douse all your sails and turn on the engine
• Turn around and run back where you came from
• Pick a new destination
• Reef and carry on
• Heave-to while you consider your options

Fog Rolls In

You're destination is nearly in sight and a thick fog bank rolls in. Now what?

Get a bearing to your destination, and then steer to miss it! Aim intentionally to one (the safer) side until you reach a depth which matches the depth of your destination. Follow the bottom contour from there.

By missing intentionally to one side you'll know which way to turn as you near shore.

If a thick fog rolls in shortly after your 9:02 fix, then a new strategy is called for on your harbor approach. Aim to miss to one side, and then follow the bottom contour to the nav aid. If visibility remains poor, then continue on the contour to a safe anchorage, anchor, and wait.

You Hear a Ship in the Fog

When underway in restricted visibility few sounds are more haunting than the low thumpa thumpa of a ships engine. It can be very hard to identify the direction of the sound. Two things to do:
• Hail on VHF channel 9 at low power. Failing that, try channel 13. Identify your location, course and speed to make sure the ship has you on radar. By all means take their suggestion if they recommend you alter course.
• Try cupping your ears and facing one side, then turn and face the other way. Do the same facing forward and aft. You may be able identify which quadrant the sound is coming from. Go the other way. If the sound remains astern and gets louder then turn hard right or left.

A Buoy is Missing

Are you sure? Better check again. Review your DR, last fix, and GPS. While you check, stay in place, or reverse course – known good water, back toward a known location. Anchored buoys tend to stay on station. 99.9% of the time the boat is in the "wrong" place, not the buoy.

Stop in safe water. Check your work every possible way. Study the chart. Consider issues like current, crossed up numbers, or out of date charts.

If the buoy is truly missing, then report it to the Coast Guard - but check again before you do that! It's much more likely that....

You're Lost

Sailing late in the afternoon through a rocky area in a swift current, you're suddenly trapped in a dense fog, as though in a white cave. The buoys disappear in the fog. Other than the blast of the foghorn on the blind lighthouse and the clang of some bell buoys somewhere nearby, there's nothing to tell you where you are. Suddenly you hear the slap of waves on rocks. You desperately look at your GPS and it is off.

What should you do? Here are some options:
• Post a lookout.
• Heave-to or anchor and double-check your position, or at the very least, slow down.
• Turn around and retrace your steps. Sail away from the rocks while checking the depth.
• Try to identify the sounds.
• Turn on the GPS.
• Check your DR plot based on your last fix
• Follow a depth contour to a nav. aid, and *buoy hop* (navigate from aid to aid) from there.

The floorboards are awash. You taste the bilge water, and it is fresh

Most likely, water is leaking from or siphoning from the fresh water tanks (or you are sailing on a lake). When heeled, water can siphon out of windward tanks. It is best to close tank and sink valves when under way.

Same thing but you taste it and it is salty.

First, be happy that it doesn't taste oily. What a bilge! Now, where is the water coming from? Most likely the head valve was left on "Flush." If not that, then check other sinks and through-hulls. You've either got siphoning, or a failed hose clamp, or a leaking fitting.

The Dinghy in Tow Flips or Swamps

Then the painter snaps from the huge load on the line. Better that than tearing the fitting off the dinghy.

Here's a great drill: Rescue the dinghy without getting anyone hurt. You'll have to stop along side, and secure a line to the dinghy. Try to lift the bow and drain the boat gradually, then lift it aboard. A block and tackle such as is used with the LifeSling may be needed.

Lost Dinghy

You go to sleep with a dinghy astern, and at dawn it is gone... Stolen no doubt. Not bad knots. No way. Just the same, if the dinghy's gone in the morning, start the search downwind. Failing that, check dinghy docks ashore, where it may have been abandoned after being "borrowed," or returned after being found, who knows...

You'll sleep better if you tie the dinghy with two painters secured to different points in the dinghy and on the boat, using good bowlines (not clove hitches). Keep it close so it won't tangle on somebody else's boat, but far enough off so it won't bang in a calm.

Entering Port

You reach your destination at night in a thick squall. The channel entrance is narrow and tricky. What should you do?

This doesn't sound like a good time to approach a tricky channel. You'll have to choose among alternatives.

- Anchor if you can find a safe, secure spot.
- Heave-to until the weather clears.
- Find another destination.
- Head off shore.
- Study the approach and make a plan with your crew. Proceed with caution, and if things turn dicey, then turn back.

Pick Up / Drop Off

It is 6 A.M., and you have scheduled a rendezvous to pick up and drop crew for 6 P.M. 25 miles upwind.

As soon as you commit yourself to be at a particular place at a particular time, the cruising gods will conspire against you. A flexible itinerary is fundamental to a relaxed and enjoyable cruise.

Out of Essentials

Short on Grey Poupon or Oreos? Long on beverages, but short on ice? Need a tool, or spare part? Ask a neighbor for help. Any excuse will do. Some crews would prefer to be left alone. Respect that. Others might be persuaded to stop by for cocktail hour in thanks for their generosity. Be sure to forget some essentials on your next cruise, and take too much of some others, so you'll have an excuse to meet your neighbors.

You've planned a weekend cruise, and you end up anchored in a storm

Enjoy. Read a good book. Play cards. Nap. A power-boater is always going somewhere. A sailor is already there.

Pull out the engine manual, and study the bleeding and filter systems. Learn how to use all the features on your GPS. Tackle a crossword puzzle. Make soup. Take another nap.

6. Conclusion

In preparing this workbook we went to cruising friends for suggestions, and asked for a list of problems cruisers experience in the real (real?) world. After offering a list with items like "air in the fuel line" and "halyard lost up the rig" and "towed dinghy swamps" Dan Neri of North Sails offered the following:

There are other problems, real problems, but problems not often addressed, perhaps because there are no good answers:

- *You've been cruising for a week, and suddenly you realize that it is Saturday, you have just started to really melt into the onboard routine but you are expected back at work on Monday. Now What?*
- *You've been on an extended cruise, but now you are back at work and can't seem to find your work ethic? "I know I left it here somewhere..."*

We're headed out now to do further research on these very real cruising problems. You should too!

Go cruising. Go prepared, and sail well. Don't postpone until the perfect moment, or you'll never go. Go. Enjoy the friendships. Relish the challenges. Send us a postcard. Go.

North U.

CD and Book Order Form

Name _____

Address _____

_____

City _____ ST _____ Zip _____

Country _____

Phone_____-_____-_____ Fax (_____-_____-_____)

e.mail _____

Boat (type/make and length) _____

Paid by: Check # _____ MC Visa

—_—_—_—·_—_—_—_·_—_—_—_·_—_—_—_

Expiration Date: _____/_____

Signature _____

Item	Qty	Price	Total
Racing **Tactics** book	___	$25	$_____
Racing **Tactics** CD-ROM	___	$40	$_____
Racing TRIM book	___	$25	$_____
Racing TRIM CD-ROM	___	$40	$_____
Race Pack (All four above)	___	$115	$_____
Match Racing eBook on CD-ROM	___	$20	$_____
Cruising Workbook (This book)	___	$20	$_____
Cruising CD-ROM	___	$40	$_____
Annapolis Book of Seamanship	___	$40	$_____
Cruise Pack (All three above)	___	$90	$_____
Weather for Sailors book	___	$25	$_____
Weather for Sailors CD-ROM	___	$40	$_____
Weather Pack (Book and CD)	___	$60	$_____
Sailor's Library (All ten items!)	___	$280	$_____

Postage and Handling* $_____

                                 Total US$ _____

* Postage per order:
US, plus Canada and Mexico: $8. ($5 *for CDs only*)
Rest of World: $20. ($5 *for CDs only*)

North U.
ORDER ON LINE AT: www.Northu.com
BY MAIL: 29 High Field Lane • Madison CT 06443-2516 USA
BY PHONE: 203 245-0727 FAX: 203 245 -0472
BY EMAIL: bill@northu.northsails.com

All Prices in US$, subject to change